PLC Programming from Novice to Professional

Learn PLC Programming with Training Videos

2nd Edition

By

Charles J. J.

Sanusi A. L.

PLC Programming from Novice to Professional

2nd Edition

ISBN: 978-9-9211-1386-0

Published in the United States

Table of Contents

About the Authors

Charles J. J. and Sanusi A. L. are control systems engineers and content creators. They have worked with many clients across business sectors, most of which are in the PLC automation industry. They have more than a decade of experience on designing control systems for industries like refineries, steel, oil, and gas from natural gas compressor stations to cold rolling steel plants.

How This Book Can Help You

This book and its supplemental training videos make up an excellent practical training program that provides the foundation for installation, configuration, activation, troubleshooting and maintenance of Allen-Bradley's PLCs (Programmable Logic Controllers) and RSLogix 500/5000 software in an industrial environment.

The 11 chapters of this book and its training videos serve as an exhaustive collection of my step-by-step tutorials on Allen-Bradley's hardware and software. It is intended to take you from being a PLC novice to a professional. If you fall in the following categories of people, you will find this program very helpful:

- Engineers
- Electricians
- Instrumentation technicians
- Automation professionals
- Graduates and students
- People with no background in PLC programming but looking to build PLC programming skills

This book is accompanied with **100+ in-depth HD training videos**. In these videos, I use a practical approach to simplify everything you need to understand to help you speed up your learning of PLCs in general, and of Allen-Bradley's PLCs specifically.

Because I assume you have little or no knowledge of PLCs, I strongly urge you to digest all the contents of this book and its supplemental training videos (over 100 episodes). This will not only help you build an in-depth knowledge of PLCs in general; it will also help you gain a lot of job skills and experience you need to be able to install and configure PLCs.

In this book I start with the fundamentals of PLCs. I went on to touch advanced topics, such as PLC networks, virtual CPU, CPU models and what their codes mean, digital input and output configurations, and so much more. The knowledge you gain from this training will put you on the path to becoming a paid professional in the field of PLCs.

The quickest way to build skills in PLC hardware and software is to use real-world scenarios and industrial applications. **The real-world scenarios and industrial applications I treat in this book and the training videos will help you learn better and faster** many of the functions and features of both the Allen-Bradley's PLC family and their software platform.

If all you use is just a PLC user manual or its help contents, you cannot become a skillful PLC programmer. That is why I have designed this training program to help you develop skills by teaching you PLC hardware configuration and programming step by step. This will give you a big head start if you have never installed or configured a PLC before. One of the questions I get asked often by a novice is, where can I get a free download of RSLogix 500 to practice?

I provide in this volume **links to a free version of the RSLogix Micro Starter Lite** (which provides essentially the same programming environment as the **RSLogix 500 Pro)** and a **free version of the RSLogix Emulate 500.** I also provide **links to download the training edition of RSLogix 5000 / Studio 5000 Logix Designer** to your system.

First ensure you create an account at RockwellAutomation.com. Once you have done that, you don't even need to have a full-blown PLC to learn, run and test your ladder logic programs. In addition to showing you how to get these important Rockwell Automation software for free and without hassle, I also demonstrate with HD training videos how to install, configure, navigate and use them to write ladder logic programs.

Finally, help/support staff are available 24/7 to help you. So, if you have questions or need further help, use the support link provided for this training. The support staff will get back to you very quickly.

How to Use My Training Videos & PLC Codes

You will find the link to download all the training videos and PLC codes at the end of each section of this book. Just use the serial number or title of the video to locate the video you'd like to watch, as mentioned in the appropriate paragraph of each chapter.

To ensure you learn very effectively and get good practice, I strongly advise you attempt the challenges and project in this book completely on your own. Try to build your own program after you've seen me build mine in each training video. Later on, you can just cross check your programs with mine. Finally, feel free to use or modify any of my PLC programs you downloaded as you wish.

Section 1: RSLogix Software & Hardware Introduction for Beginners

Chapter 1. Software Foundation: Getting You Started with PLC Programming

In this chapter, you will learn how you can download for free the necessary tools you need to start a PLC programming on your machine without the requirements of expensive hardware. The tools I am talking about are **RSLogix Micro Starter Lite** (the free version of RSLogix 500), **RSLogix 500 Emulate** and **RSLinx Classic Lite**.

That is pretty much the entire kit you might need in order to program within your own machine and get started with some of the instructions I will been teaching in my videos. If you have not already done so, go to Chapter 6 right now to get the download link and download the tutorial videos.

1.1. How to Download Free RSLogix Micro Starter Lite Package from Rockwell Automation

Here's the link to begin downloading your package: https://compatibility.rockwellautomation.com/.

Now that you have downloaded all your videos, watch the following video:

Video 1: How to download RSLogix package

On the Rockwell Automation site, I recommend going to the "Managed Download" route because it is secure. It verifies the file integrity. However, you can use the "Direct Download" if you like, especially if the Rockwell servers are having a hard time.

1.2. How to Launch RSLogix 500 for Creating PLC Programs

After downloading those three tools mentioned in section 1.1, the next process is of course, to install them. This is very simple to do. All you have to do is double-click the shortcuts and pretty much follow the installation instructions. The RSLogix Micro Starter Lite (the free version of RSLogix 500) is the primary tool you will create your ladder logics in.

You will find MicroLogix 1000 and 1100 in the **RSLogix Micro Starter Lite** package that Allen-Bradley offers you as a free download. This RSLogix Micro Starter Lite has essentially the same programming environment as RSLogix 500. So, it is ok if you are unable to buy RSLogix 500.

In addition, Allen-Bradley also offers **RSLogix Emulate 500** for free so that you don't even need a PLC to run and test your ladder logic diagrams. RSLogix Emulate 500 can be used to simulate a real PLC.

Now I will give you a hassle-free way to download the two packages for free, and also show you how to set them up. Below are the direct download links.

1. RSLogix Micro Starter Lite, Version 8.10: 4.11.00-RSLinxLite-A-DVD (or use http://bit.ly/2Q7Mk5Z).

2. RSLogix Emulate 500, Version 6.00: 6.00.00-RSLogixEmulate500-CD (or use http://bit.ly/34D0YYh).

All these toolkits that are going to be installed in your computer. Watch the following video to learn how to install and launch RSLogix 500 software step by step if that is what you use:

Video 2: Launching RSLogix500 and Creating a First PLC Program

1.2.1. How to Install & Configure Your RSLogix Micro Starter Lite Platform

RSLinx is the software RSLogix will use to communicate with your PLC or in your emulator if you don't have a PLC.

Once you've downloaded the RSLogix Micro Starter Lite software, first extract it into a folder with your Winzip software. Then install RSLinx by double-clicking on the **setup** file in the folder. Allow the installation to complete. Restart your computer. Fig. 1.2.1.1. is a screenshot of the progress in the installation of RSLinx.

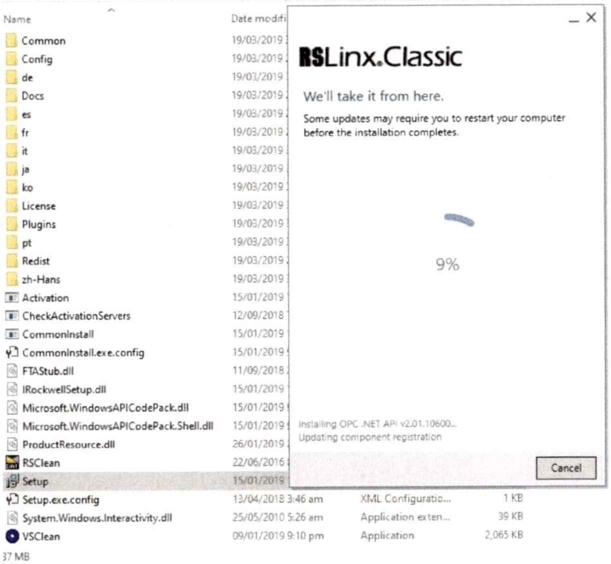

Fig.1.2.1.1. Screenshot of installation progress (9%) of RSLinx. The *Setup* file was double-clicked to start.

Next extract your RSLogix Emulate 500 in a new folder and then install it. Fig. 1.2.1.2 is a screenshot of the beginning of the installation.

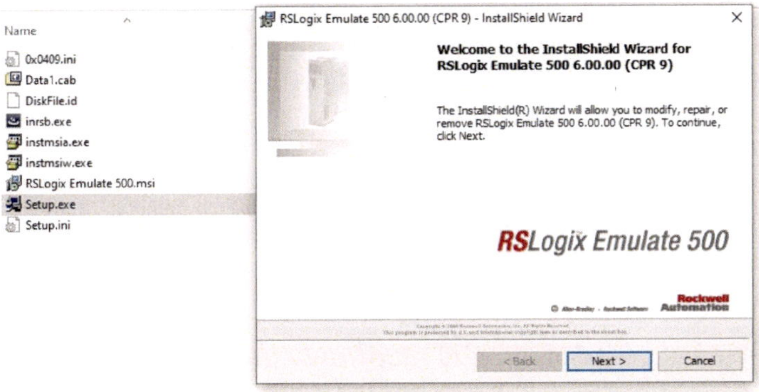

Fig.1.2.1.2. Screenshot of the beginning of the installation of RSLogix Emulate 500. The *Setup* file was double-clicked to start.

1.2.2 How to Configure RSLinx

RSLinx is the software that RSLogix uses to communicate with a PLC or the emulator if that is what you use. To start, look for RSLinx app in the list of programs/apps in your computer. Run it by following this path (assuming you use Windows 7):

START > All Programs > Rockwell Software > RSLinx > RSLinx Classic.

For windows 8.1 or 10, Fig. 1.2.2.1 shows the location of RSLinx.

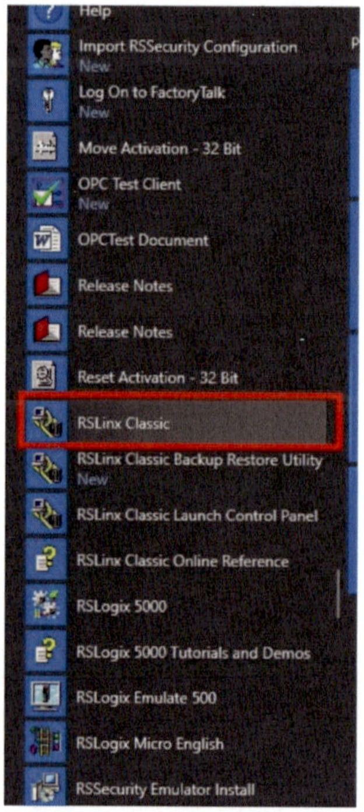

Fig. 1.2.2.1. Screenshot of RSLinx Classic app in Windows 10.

The **RSLinx Classic Lite** window opens. For now, ignore the default **Link Gateways, Ethernet driver** that shows in the window. Follow the following steps to create a new driver:

1. Open the **Communications** menu to reveal a few options such as "Configure Drivers". See Fig. 1.2.2.2.

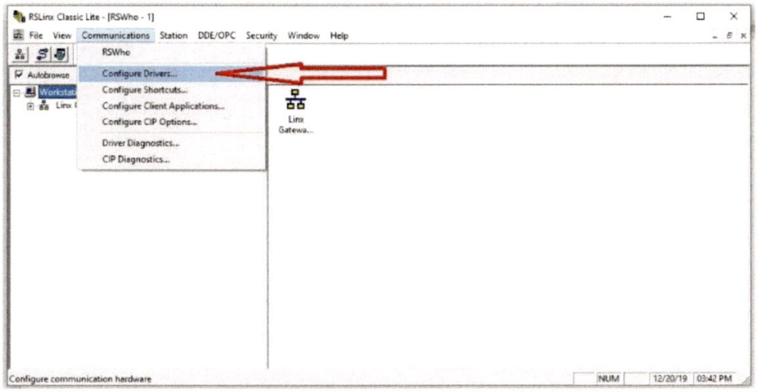

Fig. 1.2.2.2. Selecting Configure Drivers in the RSLinx Classic Lite window

2. Select **Configure Drivers**. A new dialog window with title "Configure Drivers" opens. See Fig. 1.2.2.3.

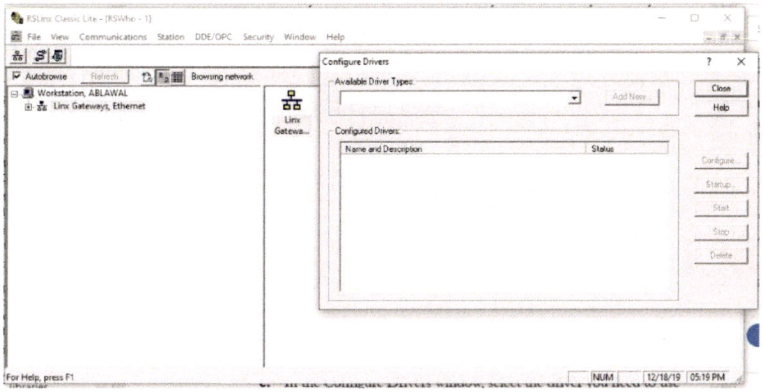

Fig. 1.2.2.3. The **Configure Drivers** window opens on top of the RSLinx Classic Lite window

3. Open the drop-down menu under the **Available Drivers Types** sub-heading. See Fig. 1.2.2.4.

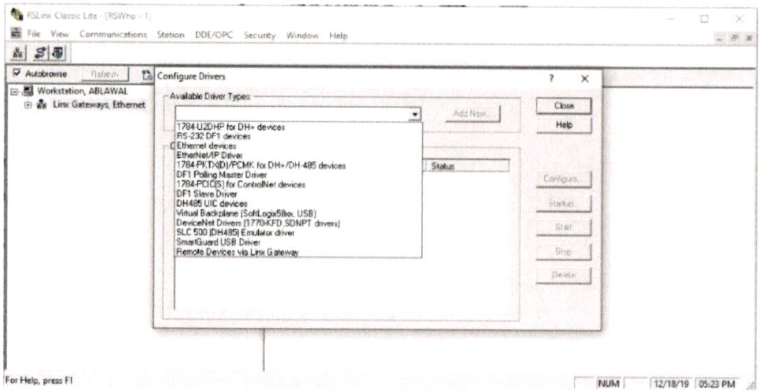

Fig. 1.2.2.4. Available driver types in the Configure Drivers window.

4. Select the **SLC 500 (DH485) Emulator driver**. Click the **Add New** button. See Fig. 1.2.2.5.

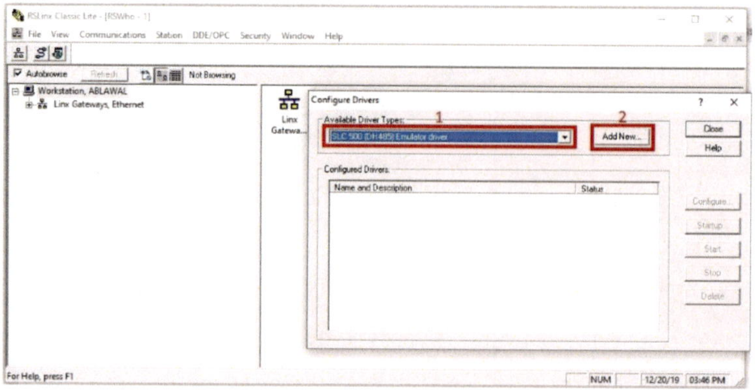

Fig. 1.2.2.5. Two steps to select and add a new driver in the Configure Drivers window.

5. You can give the driver a name, but I just leave it at the default of **EMU500-1**. See Fig. 1.2.2.6.

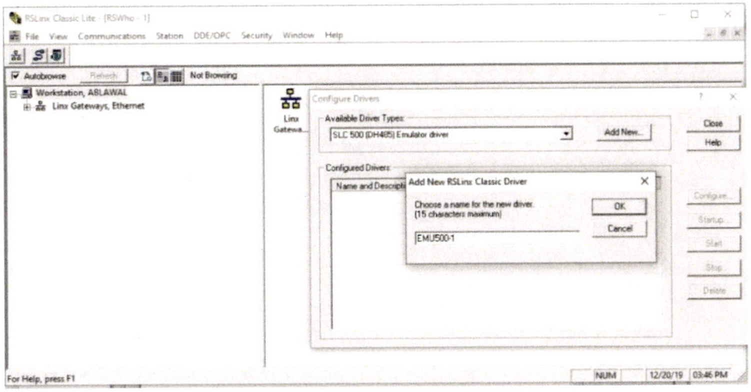

Fig. 1.2.2.6. A new driver **EMU500-1** is added in the Configure Drivers window.

6. When you click the **OK** button, a smaller dialog box titled **Configure SLC 500 (DH485) Emulator Driver** pops up, asking for the **Station Number** and **Station Name**. See Fig. 1.2.2.7.

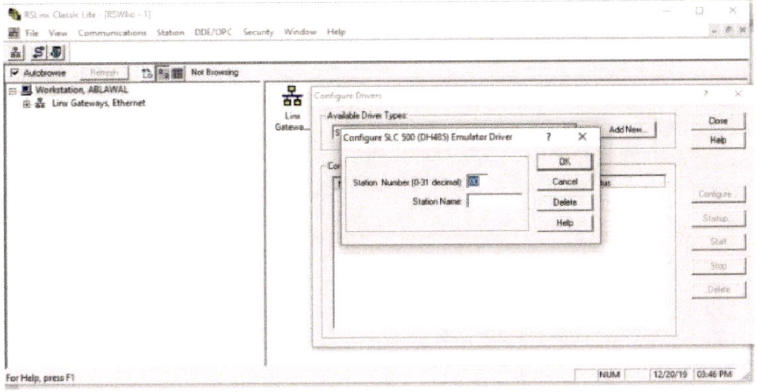

Fig. 1.2.2.7. A new dialog box asking for station number and station name in the Configure SLC (DH485) Emulator Drivers window.

7. Leave the **Station Number** as 00 and the **Station Name** empty. Click **OK**. Your new driver should now be running as shown in Fig. 1.2.2.8.

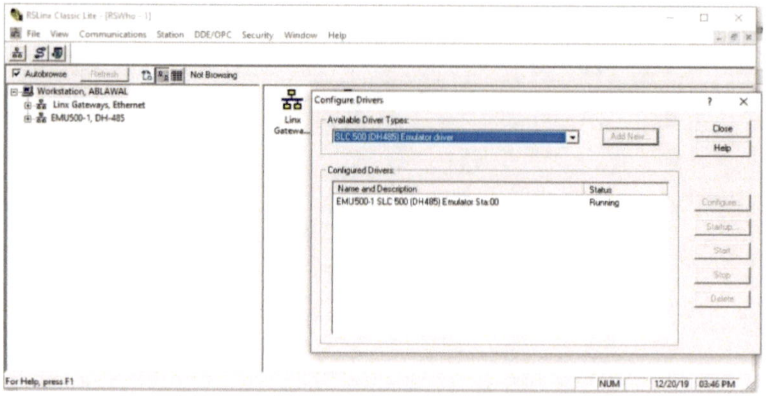

Fig. 1.2.2.8. The new driver is now running

You are now ready to go.

8. Close the **Configure Drivers** window. You will see your new **EMU500-1 DH485** driver inside the RSLinx Classic window. See Fig. 1.2.2.9.

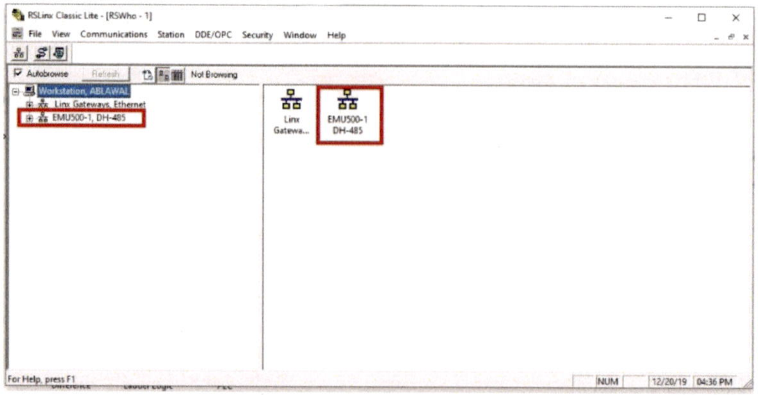

Fig. 1.2.2.9. The new **EMU500-1 DH485** driver appears in the window.

9. Now Close the **RSLinx** windows. The RSLinx is now running in the background and you'll *probably* see its little icon in the system tray. See Fig. 1.2.2.10.

Fig. 1.2.2.10. RSLinx icon in the system tray

1.3. Getting Started with RSLogix500 on an Actual MicroLogix 1100 PLC

In this section, I want to show you how to work with a real processor. So, you should study this section and watch the video if you are looking to invest and perhaps purchase one of those Allen-Bradley controllers from eBay or other sources. I'm going to show you step by step how to work with an out-of-the box controller (MicroLogix 1100), how to set an IP address, how to download your first program, and essentially go online with the processor so that you can get started programming on a **real PLC**, not just a simulation, which you have seen in the last couple of videos.

Let's learn about the hardware setup like that shown in Fig. 1.3.1.

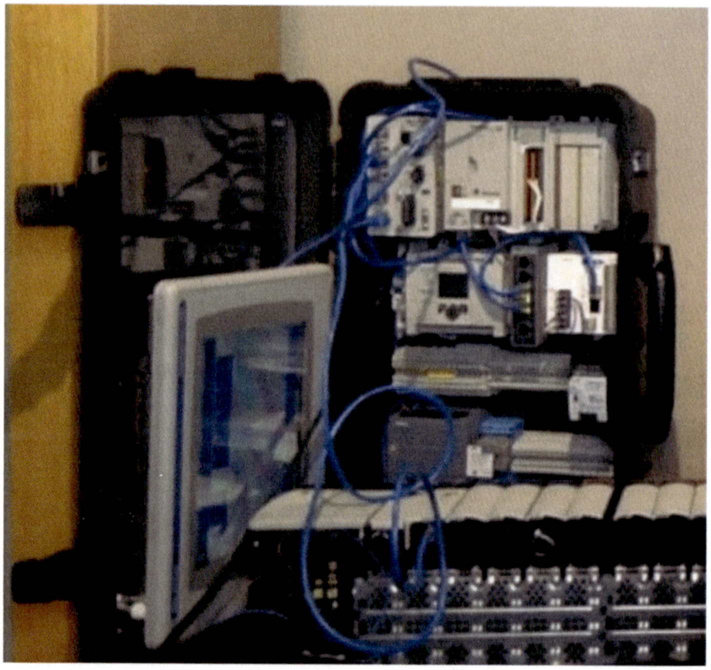

Fig. 1.3.1. My MicroLogix 1100 Controller with BOOTP/DHCP Ethernet Cable

This controller comes with essentially a DHCP or BootP method of acquiring an IP address. This means that if you don't have a server that gives it an IP, then it's essentially going to wait for you to give it an IP address. In order to figure this out and give it a static IP address, we need to do a couple of little things first.

But first and foremost, know that my controller is connected through an ethernet cable from my laptop to the PLC. It goes through a little switch sitting in the top left corner of the controller's box before it connects to the PLC. This enables a direct communication. The ethernet cable is plugged in the back of the PLC. Now watch the video to learn how to configure your adapter.

Video 3: Getting Started with RSLogix500 on an Actual MicroLogix 1100 PLC

Chapter 2. Hardware Foundation: ControlLogix, CompactLogix, MicroLogix & Other Control Systems Hardware

2.1. Introduction to PLC Hardware

In this chapter, I treat the foundation of the hardware. So, we're going to be talking primarily about the different hardware that you would find in control systems. The main pieces of hardware are of course going to be the PLCs.

Allen-Bradley has a couple of different platforms, which we're going to be looking at. Fig. 2.1 shows an example of a very large rack of ControlLogix system.

Fig 2.1: My rack of ControlLogix system

I have the PLC, as well as many diverse different cards. We're going to take a look at each and every card. We're also going to talk about how to wire them up and discuss the different advantages of the particular platform, as well as how it pertains to a control system overall.

We're also going to be talking about several other platforms, such as the Allen-Bradley families of the CompactLogix. I will also show you a Micrologic series processor such as the 1100 series PLC, which is one of the most recommended MicroLogix PLCs for beginners because of its ease of use, accessibility, and affordability.

We're going to be exploring each individual platform in a lot of detail. We're going to be talking about the different inputs and outputs, and what it means to have an analog versus a digital signal.

We're also going to discuss the different push buttons, safety circuits, and how to manage the platforms in general. You are going to learn how the hardware fits together into this entire family. Of course, you're going to have a lot of practice exercises on which you can rely in order to strengthen your own knowledge of the systems.

This will help you practice as you go. So, try to attempt some of these exercises in your own spare time. Without any further delay, let's jump into the next section to look at the ControlLogix series.

2.2. ControlLogix - Chassis & Power Supply Overview

As I mentioned in the previous section, we're going to look at the ControlLogix platform from Allen-Bradley and our focus today is going to be on the chassis, which I have sitting on my desk right now, as well as the power supply sitting right next to it. See Fig. 2.2.1.

Fig. 2.2.1: ControlLogix Chassis (right) and a DC Power Supply (left)

I'm going to explain step-by-step what you should expect in the field, how you can find them in different panels, and how you need to assemble them.

You will learn pretty much everything you need to know from a high-level perspective, as well as diving deep into little details.

Watch the next video now.

Video 4: ControlLogix - Chassis and Power Supply Overview

2.3. ControlLogix - PLC 1756-L61 Overview

In this section we're going to talk about the most important card of our platform. This is of course, the programmable logic controller 5561 processor (PLC 1756-L61), which is one of the more commonly known processors used within the industry right now. See Fig. 2.3.1.

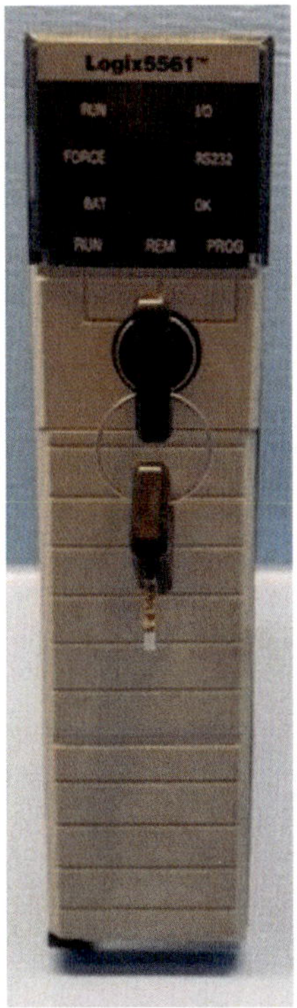

Fig. 2.3.1: Logix5561 Processor (PLC 1756-L61)

This is a ControlLogix platform processor, and it goes inside of the chassis as you saw in the previous video. We're going to take a very deep dive into this controller, that we will use the same approach of a top-down view of the card.

I'm going to be walking you through every single feature that you should be familiar with, such as how it operates, how to impact your system, as well as what kind of a role it plays within the infrastructure itself. So, without any further delay, let's jump right into the Logix5561 processor.

Video 5: ControlLogix - PLC 1756-L61 Overview

2.4. ControlLogix - 1756-EN2T EtherNet_IP Adapter Overview

In this section we will talk about a very critical module for the ControlLogix platform. This module is known as the 1756-EN2T, otherwise known as the ethernet communication module. It allows the user (or the controller itself) to talk to other devices over ethernet IP, and essentially send or receive data that is going to be critical for the process. See Fig. 2.4.1.

Fig. 2.4: ControlLogix module 1756-EN2T

It also allows you to go online from, for example, a plant network to a different location through ethernet IP in order to troubleshoot a program. It has other kinds of features that you might want to use. The thing that you need to know is that there's a couple of different flavors. So just like with the L61 processor, where you now have the L71 and L81 processors, we now have different flavors of these ethernet cards as well. An example is 1756-EN3TR.

As years pass by, we get more and more features. This is EN2T card, but there is also EN2TR card with more features added. However, at the core level, the cards are still the same. You can still see the EN2T card used heavily in the PLC industry.

In the next video we're going to be taking a very deep look at the EN2T card, but essentially what you need to know as well as it provides all your functionality, for example, to communicate from the user as well as other devices over Ethernet IP.

Video 6: ControlLogix - 1756-EN2T EtherNet_IP Adapter Overview

2.5. ControlLogix - Input & Output Modules - Analog, Digital, AC & DC Overview

In this section we're going to talk about the different input and output modules within the ControlLogix family. We will look at four distinct modules which are going to be used to tie in all your instrumentation, such as sensors and valves.

Some examples are position valves, temperature sensors, and flow meters. Things like these are going to be connected to your input or output modules, of course depending on whether it is an input or an output.

Therefore, in practice, you're going to have an array of different cards that are going to be applicable to only certain devices. For example, in Fig. 2.5.1, we have an analog input card, as you can see by the labeling at the top.

Fig. 2.5.1: An Analogue Input Module 1756 IF8

Fig. 2.5.2 shows an example of a digital input card.

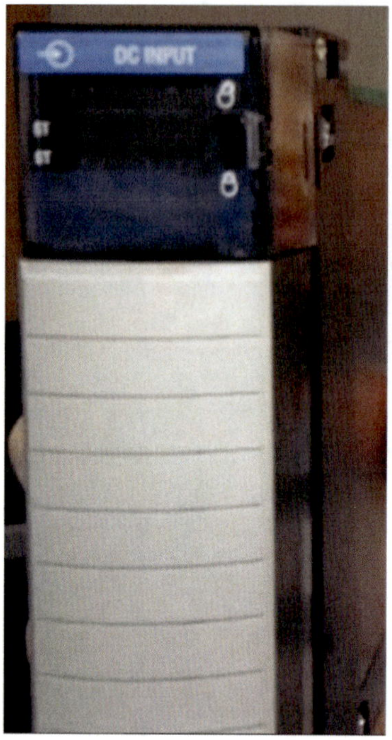

Fig. 2.5.2: A Digital Input Module 1756 IB16

Fig. 2.5.3 shows an example of an analogue output card.

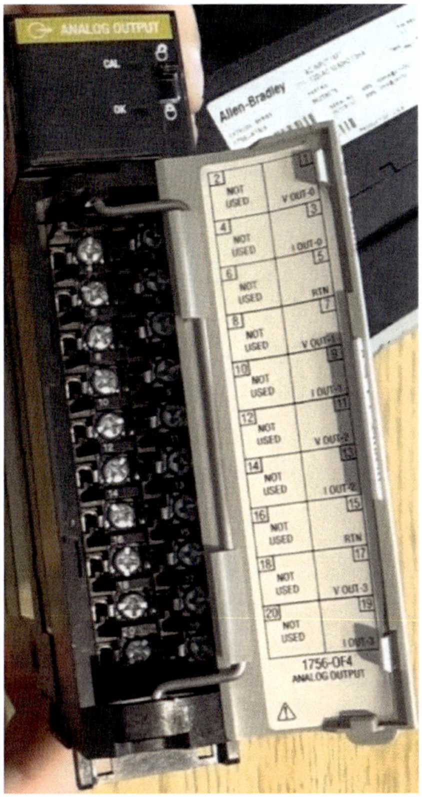

Fig. 2.5.3: An analogue output Module 1756-OF4

You can see more examples of these cards at Indiamart.com.

These cards are similar in their functionality, though may look different in appearance. All the cards look essentially the same, like you've seen with the PLC as well as the ethernet card. However, they do have different features and different capabilities because they are essentially designed for different instrumentation.

You should be able to not only understand what they do, but also have a way to select which cards you need, depending on what kind of a control system you're going to be designing as an engineer or even a plant technician.

So, it is very important to be familiar, at least with the most common cards that are used in control systems. Of course, as you develop your knowledge, you are going to be trying to figure out which cards are the best, or which cards are going to be more applicable to your particular case. So, you will get more and more familiar with these different input and output cards.

In the next video but we will look at them a little bit more closely.

Video 7: ControlLogix - Input and Output Modules - Analog, Digital, AC and DC Overview

2.6. ControlLogix - Chassis Assembly & Power ON - Analog, Digital, AC & DC Overview

In this last section on ControlLogix hardware, we're going to be putting everything together. Now that we've learned about all the modules that we need to be familiar with, we're going to start plugging them into the back plane.

If you remember from Section 2.2 which featured the backplane, we talked about putting all the modules in their slots. So, in the following video I have my PLC plugged into the slot zero. In case you're wondering if there's any reason behind where you put in your modules, there is no wrong answer. See Fig. 2.6.1.

Fig. 2.6.1: ControlLogix Chassis Assembly

There are essentially no problems with putting those modules in whatever order you desire. However, you do want to keep some consistency. Here's the video where I explain everything in detail:

Video 8: ControlLogix - Chassis Assembly and Power ON Analog, Digital, AC, DC Overview

2.7. CompactLogix - Platform Overview - Analog, Digital, AC & DC Overview

Welcome to this new section, in which we're going to be discussing CompactLogix. This platform is slightly smaller, and it is typically used for smaller equipment and standalone machines, as opposed to ControlLogix which is used for plant-wide equipment.

Now that's the main difference, but ultimately it comes down to cost. It is a much cheaper alternative. However, it still runs RSLogix, as well as Studio 5000, and it's still made up of the same or similar modules that you have seen before.

Therefore, we're not going to spend a lot of time on this platform, so I'm going to get straight into it. You're going to see everything in one single video that follows, and we're going to be covering the PLC as well as some expansion cards that we can use with it.

Video 9: CompactLogix Platform Overview - Analog, Digital, AC and DC Overview

Chapter 3. Software Development in RSLogix 5000 & Studio 5000 Ladder Logic

3.1. Introduction to Software Development

This is the introduction to the programming section. We have a lot of very important sections ahead of us, where we will cover all the instructions that you need to know, such as XIC, XIO, PID loops and so on.

Before we jump into the programming aspect itself, I want to take a step back and essentially go through certain things which are required before we actually go online and program a PLC. Those things could be upgrading or even downgrading your firmware levels, depending on which software you're using to program that PLC. It could also be how to go online with the PLC, setting your IP addresses right, making sure that the network is correct, and so on.

All of these are essentially a part of the programming section, and we are going to spend a little bit of time. Like I said, I'm going to show you exactly what you need to do to get the latest firmware.

You need to know which firmware is compatible with your particular processor, because at the end of the day, you do need to go through those steps every single time that you are working with an unfamiliar processor.

It's very important to know, not only for the purpose of a job interview, but also because once you start working at a plant or a facility which uses these processes, you essentially need to know what to do in order to go online with them and be able to program them.

Like I mentioned a bit earlier, we're going to take a deeper dive into each one of those programming instructions. I'm going to cover them step-by-step because it's going to be aimed towards someone who's not familiar with them.

But if *you* are more of an advanced user, feel free to skip the sections that you're already knowledgeable in and go directly to the instructions that you're interested in, such as the advanced techniques in a software programming.

3.2. How to Download Firmware from Rockwell Compatibility Center

In this section we're going to look at the product compatibility center from Rockwell Automation. This is the best tool which I strongly advise you use to determine the right firmware levels of your processor or other parts.

Before you dive into the video, I just want to briefly mention that firmware levels are a very big deal because, depending on the products you're using with your PLCs, **the firmware levels need to match or need to be compatible with each other**.

Not only do you have to worry about the firmware level of the PLC itself, you also need to worry about the firmware level of each and every component that will interact with your PLC system.

Here’s the link to check out and begin downloading the right firmware revision of your processor: https://compatibility.rockwellautomation.com/.

Here’s the video where you can learn the detail:

Video 10: How to Download Firmware from Rockwell Compatibility Center

3.3. RSLinx Classic Setup & Drivers

RSLinx Classic is very important tool to know if you're planning to work with a PLC system. In the next video, you will learn exactly how to configure this tool in order to be able to communicate with your processors, and to see all the important information and configuration settings for your particular devices.

Watch the video below to learn how to create a new driver and figure out revision level through your RSLinx Classic.

Video 11: RSLinx Classic Setup and Drivers - Analog, Digital, AC and DC Overview

3.4. RSLogix 5000 & Studio 5000 Firmware-Software Revisions

In this section we will continue to work on our software and firmware revisions for our PLCs. There is an extremely important thing I want to mention here.

It is the fact that **your RSLogix 5000 (or Studio 5000) installation needs to match the firmware revision that you have on your PLC** in order to be compatible and be able to program. Their major revisions do need to match exactly.

Sometimes their minor revisions also need to match. The easiest way to find out the version of your RSLogix to work with is to launch it on your PC and then look it up at the bottom of the window during the startup process. An example is shown in Fig. 3.4.1 where the version is indicated at the bottom of the startup window.

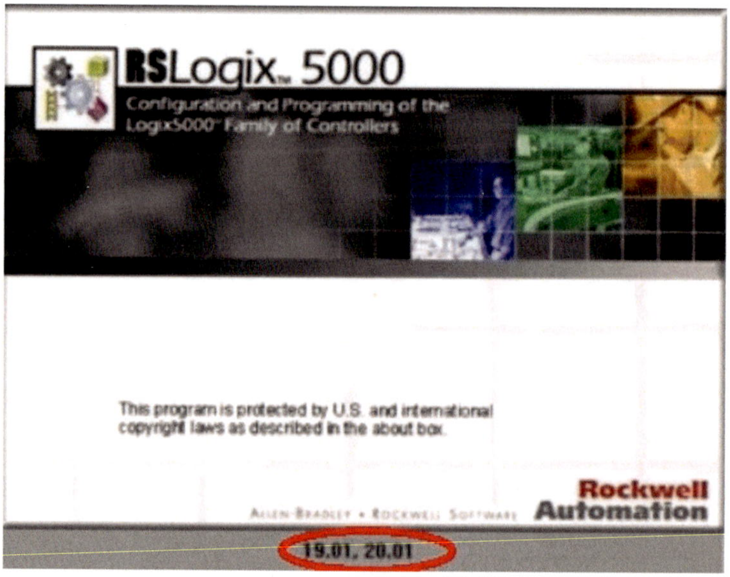

Fig. 3.4.1: How to check the version of RSLogix 5000

A very common question that is often asked by new PLC programmers is can you work with a higher revision of RSLogix with a processor of lower revision? The simple answer is no. I highly recommend you watch this video now to learn more:

Video 12: RSLogix 5000 and Studio 5000 Firmware_Software Revisions

3.5. PLC Firmware Flashing, First Program Creation & Online Connection

In this section we are going to work with RSLogix 5000 version 19.11. Currently it is running a 1756-L61 PLC, which is a. In the last section, I re-flashed the firmware of my PLC back to version 20.11.

So, in this section I'm going to briefly show you how to make this change in your own software too if you have either a different application or if you have an existing application that needs to match your controller type.

Go over to the next video now to learn how to do this, and to also understand the RSLogix 5000 platform in general. Finally, after watching this video you should be able to write your first simple PLC program:

Video 13: First PLC Program Creation and PLC Online Connection

3.6. Getting Started with the Basics: Routines in RSLogix 5000

In this section, we're going to establish our first level of rungs to program within the RSLogix 5000 series. There are two ways to program your controllers: the first way is online, and the second way is offline.

In online programming you are connected live with the processor, while in the offline programing, you are only programming on your PC without making changes to the live environment. There're a couple different reasons why you would want to do either one.

For example, it's always much easier to program offline because you have much fewer restrictions. Also, you can create new tags willingly or delete tags. Furthermore, programming online could create some problems. For example, you might get errors and you will then need to recompile your program.

So, we're going to start off with is offline using the “Rem Run” button on the RSLogix 5000 platform. Learn more in the following programming tutorial:

Video 14: Getting Started with Routines and RSLogix 5000 Basics

3.7. How to Build Rungs & Understand the Principles of Ladder Logic Structures

In this section we're going to talk about the basic principles and structure of ladder logic programming. You will learn how to create fundamental AND/OR statements, how to create multilevel rungs, how to use branches on your rungs, and essentially how the rungs operate from a general perspective.

So, I want you to pay very close attention as you watch this episode.

Video 15: Building Rungs and Understanding the Principles of Ladder Logic Structures

3.8. Good Practices of Structuring Rungs: Leaving Rung & Tag Level Comments

As promised in the previous section, we're going to look at proper structures of rungs, tags, and descriptions. In order to do that, I will give you an overview of a previous program I wrote myself.

I will walk you through some of the design choices, but I because also don't want you to be overwhelmed, I'm going to show you some very simple code. I want you to pay attention so you can understand the way things are laid out.

So, as you move forward, you will learn some of the different instructions that you can use to start practicing on your own. Let’s get started:

Video 16: Good Practices of Structuring Rungs, Leaving Rung and Tag Level Comments

3.9. Programming Instructions: XIC & XIO - Examine If Closed & Examine If Opened

This section is a quick tutorial on XIC (Examine If Closed) and XIO (Examine If Open) instructions. These are two of the most fundamental yet sometimes misunderstood instructions in the ladder logic environment. This is something that is common to a lot of PLCs.

Let's take a look at the definitions of these generic terms and their symbols that are used when we talk about PLCs:

--] [--

Normally Open Contact or **NO (**called **XIC** - eXamine **I**f **C**losed - by Allen-Bradley**)**: When a NO contact is used with a hardwired input, the instruction is **OFF** or **LOW** until a voltage is applied to the input. The bit address then goes **ON** or **HIGH**, that is, the instruction becomes "true." It works the same way when it has the same address as an internal coil, except that the coil must be turned on by logic in the program.

--] / [--

Normally Closed Contact or **NC (**called **XIO** – e**X**amine **I**f **O**pen - by Allen-Bradley**)**: It is an inverted normally open contact. When a NC contact is used with a hardwired input, this instruction is **ON** or **HIGH** (that is "true") until there is a voltage applied to the input. It then goes **OFF** or **LOW** and becomes "false". It can also be used with an internal coil, becoming true when the coil is off and becoming false when the coil is on.

In any typical PLC program, you would see between 90 to 95% of XICs and XIOs used. So, it is extremely important to understand how they work. They are not that complicated as you will learn in the following video tutorial.

Video 17: PLC Programming Instructions - XIC and XIO

3.10. Programming Instructions: OTE - Output Energize

In the previous video, we looked at XIC and XIO instructions within our ladder logic environment. In this section, we're going to continue with basic instructions. We will look at the OTE (Output Energize) instruction, which is extremely critical because it is used all over the PLC industry. A single bit of data is energized by an OTE instruction if the input leading to it is high or true.

Let’s take a look at the definitions of this generic term and its symbol that is used when we talk about PLCs:

-( )-

Output Coil (called **OTE** – **Ou**tput **E**nergize by Allen-Bradley): When an output coil is used with a hardwired output, this function is **OFF** until the logic in the program allows it to turn on. It then becomes “true” and will energize the device that is wired to the respective output. If it is used as an internal coil, it will toggle the instructions associated with it. That is, it will close a normally open instruction and open a normally closed instruction.

Without any further delay, let's jump straight into the tutorial.

Video 18: PLC Programming Instructions - OTE - Output Energize

3.11. Practical Challenge 1: 3 Buttons & 3 Lights Control System (Output Latch - OTL & Output Unlatch - OTU)

Welcome to your first project challenge. This challenge is a small piece of code that we're going to make together. You will have an opportunity to practice solving a real-life problem that you might encounter in the field.

Let’s take a look at the definitions of these two generic terms and their symbols that are used when we talk about PLCs:

–< L >–

Output Latch (**OTL**): An OTL instruction will force a single bit of logic into a HIGH state if all the conditions leading to it evaluate to true. It's a powerful basic instruction but can cause you a lot of grief if you use it improperly or too frequently.

The OTL instruction is often found on the right side of a ladder logic rung. It switches a bit to a HIGH state once it executes. Unlike the OTE Instruction, the **OTL will never turn the bit LOW**. In order to make that happen, you can leverage other instructions. Another instruction most commonly paired with OTL is an **OTU** (Output Unlatch).

–< U >–

Output Unlatch (**OTU**): The OTU instruction will set a bit to a LOW state if all the conditions leading to it evaluate to true. This instruction is one of the fundamental instructions for PLCs. However, it should be used with caution. This instruction will be found on the right side of a ladder logic structure and turn a bit to a LOW state if the preceding instructions evaluate to true.

Now, there's a couple of things that I want you to keep in mind before you jump into this challenge. Since we have not looked at some of the more advanced topics such as how to develop your own screens on the **HMI** (Human Machine Interface), I want you to ignore that for now, and just look at the very simple inputs and outputs shown in Fig. 3.11.1:

Fig. 3.11.1: 3 Buttons and 3 Lights Control System

The second thing that I want to point out is that I'm going to be doing this exercise in RSLogix 5000, instead of RSLogix 500. Now, the point that I want you to take away from this is that ladder logic can be written in different languages. So not only do you not need to stick to RSLogix 500 or even 5000, you can also migrate to Studio 5000, or even go with a Siemens PLC. PLC concepts are applicable to different platforms.

Without any further delay, let's take a look at the training in the following video:

Video 19: Product Challenge 1 - 3 Buttons and 3 Lights Control System

3.12. Programming Instruction: One Short (ONS)

Here we're going to talk about the One Shot instruction, commonly known as the **ONS**. This instruction allows a rung to execute only once if the data leading to it is TRUE. ONS will not allow the rung to evaluate again until the input is set back to LOW and then to HIGH.

The following video is a little demonstration of ONS I set up for you on my PLC:

Video 20: PLC Programming Instructions - ONS - One Shot

3.13. Programming Instruction: One Shot Rising (OSR) & One Shot Falling (OSF)

In the previous video, we talked about the One Shot instruction as a very useful instruction that activates at the rising edge of a condition. Now in this section, we're going to talk about two sister instruction to the One Shot.

The first one is are called the OSR (One Shot Rising) and the second one is called OSF (One Shot falling). These two instructions are very similar in nature but provide a little bit more flexibility to the programmer. They allow you to create a little bit more complicated routines and kind of save on how many bits you are using.

I have a very good demo in the following video:

Video 21: PLC Programming Instructions - OSR and OSF

3.14. Programming Instructions: Timer ON (TON)

Here we're going to look at a timer instruction called the Timer On delay. It's a very important instruction and is used quite frequently for anything related to timing sequences or any sort of time events. Essentially, it is used to delay the start of a process or a machine for a certain period of time.

The following demo is a fairly straightforward example to understand. Check it out now:

Video 22: PLC Programming Instructions - TON - Timer ON

3.15. Programming Instructions: Timer OFF (TOF)

In this section, we're going to talk about the Timer Off delay instruction, commonly called the TOF instruction. It is a PLC programming instruction which is used to switch off the output or a system after a set period of time.

In the previous video, I spoke about the Timer On delay, which is the more common counterpart that I currently see in my PLC programs and all over the PLC industry.

But the Timer Off delay certainly has its own applications. In the following tutorial, we're going to be look at one of them.

Video 23: PLC Programming Instructions - TOF - Timer OFF

3.16. Practical Challenge 2: Tank Filling Mini-Project (HMI-PLC Development, Part 1)

This is going to be your second practical challenge (a tank filling mini project) and we're going to go a little bit more in depth. That is, our system is going to be a little more complex. This is not going to be just three rungs of logic like we did in the previous challenge.

What I have for you in the next video is essentially a tank-filling system which is illustrated in Fig. 3.16.1. I hope you will be able to apply some of what you learned in the previous sections to this challenge.

Fig. 3.16.1: A tank filling system

Video 24: Product Challenge 2 - Tank Filling Mini-Project HMI_PLC Development Part 1

3.17. Practical Challenge 2: Tank Filling Mini-Project (HMI-PLC Development, Part 2)

This is the second part of our challenge, the tank filling mini project. In the previous video, we wrote a couple of rungs of logics, which are essentially controlling the starting and stopping sequence of our tank.

In this section, we're going to delve into the HMI and start automating certain things. So, what I'm going to do first is work with the start and the stop buttons for both filling and discharge stations, and then link thme properly through the HMI.

Here's the detail in this demo:

Video 25: Product Challenge 2 - Tank Filling Mini-Project HMI_PLC Development Part 2

3.18. Practical Challenge 2: Tank Filling Mini-Project (HMI-PLC Development, Part 3)

This is the third part of our challenge, the tank filling mini project. In this section we continue to build our HMI, create HMI tags and continue to automate things. Here's the link to the next demo:

Video 26: Challenge 2 - Tank Filling Mini-Project HMI_PLC Development Part 3

3.19. Practical Challenge 2: Tank Filling Mini-Project (HMI-PLC Development, Part 4)

In this section, we're going to finalize everything on our control system. We're going to make the HMI work and display the statuses of our "Filling" and "Discharge" stations. In addition, we're going to display the really cool bar in the middle of the tank. Without further delay, let's do that right now:

Video 27: Challenge 2 - Tank Filling Mini-Project HMI_PLC Development Part 4

3.20. Programming Instructions: Count UP (CTU)

Now we will move forward with our programming instruction tutorials. In this section we're going talk about the Count Up, which is more commonly referred to as the CTU instruction. It's an extremely useful instruction in RSLogix and many other types of ladder logic programming.

It is often used to count up a number of events. For example, think of a proximity switch which counts the number of products that go by. Another example is a laser sensor that is looking at different products going up or down on a conveyor.

Yet another example is a counter that counts the number of button presses on a HMI to keep track of events on your system. So CTU is an extremely versatile instruction, and it is used in very many different applications.

Now let's start off with a demo right away:

Video 28: PLC Programming Instructions - CTU - Count UP

3.21. Programming Instructions: Count Down (CTD)

In this section of our PLC programming we're going to look at yet another ladder logic instruction within the RS Studio 5000 environment. The instruction I'm referring to here is the Count Down or CTD. It is very similar to the Count UP (CTU) instruction.

I have a couple examples for you just like in the previous episode. See them in the video below:

Video 29: PLC Programming Instructions - CTD - Count Down

3.22. Programming Instructions: Retentive Time ON (RTO)

Now I want to talk about the Retentive Timer On instruction, which is more commonly known as the RTO. It is used for storing or holding a set (accumulated) time. RTO is used in the case when there is a change in the rung state, power loss, or any interruption in the system.

As the name would suggest, it's an instruction that is used within the timer family, but it's quite different than the Timer On delay and the Timer Off delay that you saw in sections 3.14 and 3.15 respectively. The next demo video will tell you exactly why.

We're also going to look at an example of how you can create your own Retentive Timer On through a Timer On delay instruction. Let's jump right into this:

Video 30: PLC Programming Instructions - RTO - Retentive Time ON

3.23. Programming Instructions: Equals (EQU) & Not Equals (NEQ)

We are continuing our journey into the exploration of different ladder logic instructions that PLC programming has to offer. Still within our Studio 5000, we are now going to look at the comparison section.

We're going to explore the Equals and the Not Equals instructions. They are very straightforward and easy to use, but they do deserve an example. So, let's dive right into them:

Video 31: PLC Programming Instructions - EQU and NEQ

3.24. Programming Instructions: Greater Than (GRT) & Less Than (LES)

These are yet another set of instructions which we are going to explore in our RSLogix or Studio 5000. There are four of them actually: Greater Than (>), Less Than (<), Greater Than or Equal to (≥), and Less Than or Equal to (≤).

They are extremely useful bits of logic and are often used in instructions for anything involving counters, timers, or anything that need to essentially calculate or compare two different integers, constants or floats.

Since these 4 instructions are extremely similar, we are going to cover them in a single video. Now let's just jump straight into the tutorial:

Video 32: PLC Programming Instructions - GRT and LES - Greater and Less

3.25. Programming Instructions: Limit Test (LIM)

In this section, we're going to explore a different compare instruction. The instruction that we're going to talk about is the Limit Test or LIM instruction, which is used to test a single integer or a float against two other different values.

LIM has a lot of different uses. I use it quite frequently in my own logics. So, I highly recommend you become very familiar with this instruction. Now, let's jump right into this instruction:

Video 33: PLC Programming Instructions - LIM - Limit Test

3.26. Programming Instructions: Addition (ADD) & Subtraction (SUB)

It's time to look at two more instructions within the **Compute/Math** sub category. They are Addition (ADD) and SUB (Subtract) instructions. These two instructions are straightforward from a mathematical standpoint. However, they do have a couple of nuances to them. So, we're going to explore them by looking at a couple of examples in this video:

Video 34: PLC Programming Instructions - ADD and SUB

3.27. Programming Instructions: Masked Equals (MEQ) - Part 1

Now let's talk about the Mask Equals or MEQ instruction. It's a fairly straightforward instruction, but there's quite a bit of detail involved in understanding the mask itself. The MEQ is something that is not used very frequently, but the mask itself is used in many different instructions. So, this is the perfect way to understand what the mask actually does if you're not familiar with bitwise operations.

In the following video, we're going to take a deep dive on every step, and I will explain to you everything you need to know about this instruction, how it is applicable to your situation, and how it can be used. Of course, you will also understand the ins and outs of how bit processing is done in this particular instruction. Now let's dive into it immediately.

Video 35: PLC Programming Instructions - MEQ - Masked Equals

3.28. Programming Instructions: Masked Equals (MEQ) - Part 2

In the previous section, I talked about the MEQ or the Mask Equals instruction. This instruction is very useful in certain specific cases. But I got some feedback that some of my students couldn't understand the instruction so I decided to do one more practical video (Part 2) on this, on where this instruction might be used.

So, I created a second part for the explanation of this instruction because I do believe that there's some value in using it. I believe in this section you will understand better how it works from a general logic standpoint, and also in certain specific applications.

We're going to understand step-by-step not only why this instruction could be applied to the example in the next video, but also some of the other ways that you can optimize your PLC programming code, make it a bit more efficient, and make it take up less space.

Without any further delay, let's jump into Studio 5000 and get started:

Video 36: PLC Programming Instructions - MEQ - Masked Equals - Part 2

3.29. Programming Instructions: Multiply (MUL) & Divide (DIV)

In this section, we're going to be looking at more instructions within the Compute/Math sub category. The two instructions we're going to learn about today are the MUL (Multiplication) and DIV (Division).

These two instructions are not very complicated from a mathematical standpoint. I'm going to show you a couple of very creative ways of using them, and also show you a couple of practical examples as well. So, let's just dive into the demonstration:

Video 37: PLC Programming Instructions - MUL and DIV - Multiply and Divide

3.30. Programming Instructions: Modulo (MOD)

Welcome to at another tutorial on different PLC instructions that you can find in RSLogix and Studio 5000, as well as many other PLC programming languages. In this section, we are going to look at the Modulo instruction.

It's a very straightforward mathematical instruction, which may or may not be familiar to you. What it allows you to do essentially is state the remainder of a division. So, in many circumstances, whenever you divide two numbers, of course, if it's not an even division you will be left with what's called a **remainder**.

For example, when 10 is divided by 3, it's going to give you an integer of 3 and then the remainder is going to be 1. Now let's dive right in the tutorial:

Video 38: PLC Programming Instructions - MOD - Modulo

3.31. Programming Instructions: Compute (CPT)

Here's another tutorial on a different instruction in RSLogix and Studio 5000. This is a very critical tutorial because we are going to explore one of the crown Jewels (so to speak) instructions within the Compute/Math segment. This instruction is of course the compute instruction or more commonly known as the CPT.

Now, this instruction is unique in the sense that it allows you to essentially combine the functionality of many other instructions that we've seen thus far into one simple expression. By simple I mean, of course, it allows you to do quite a bit of functionality within one block of your program.

Now, you of course don't want to abuse that power and you want to keep them as concise as possible, but we'll talk about that a bit later in the book. For now, let's jump right into the next video:

Video 39: PLC Programming Instructions - CPT - Compute

3.32. Programming Instructions: Bit Shift Left (BSL) in RSLogix 500

Welcome to yet another tutorial video in which we're going to explore a very demanded instruction, that is the Shift Left (BSL). In the video, I have a very basic example set up with a couple of rungs.

When you study it, try to see if you can make some sense of this seemingly intriguing instruction. I will walk you through every step of the way so that you feel very confident in how you use it in your code. That is because it is something you might see in production PLCs, and is used quite frequently for rejects and things of that nature.

Video 40: PLC Programming Instructions - BSL - Bit Shift Left

3.33. Programming Instructions: Bit Shift Right (BSR) in RSLogix 500

Welcome back to yet another tutorial on bit shift registers. In the example in the following video, you will see the only modification I've made is changing the Bit Shift Left that you saw in the previous video to Bit Shift Right.

The reason I did the Bit Shift Left first is mainly because it is a bit more intuitive to understand. You have an integer that is growing, and you can see the bits shifting all the way to the end of that integer.

However, with the Bit Shift Right, that's no longer the case. It becomes a little bit more confusing just because of the way things shift, and you start having things appear in the middle of your integer. Watch the tutorial now to understand exactly what I mean:

Video 41: PLC Programming Instructions - BSR - Bit Shift Right

Chapter 4. Advanced PLC Programming Topics

In this chapter we will look at some advanced PLC programming topics, such as loop development, PLC forces and I/O buffering. If you have studied the previous tutorials and practiced along well enough, you should not find it difficult to understand the topics we will discuss in this chapter.

4.1. PLC Programming Instructions: FOR Loop Development

This first section is a very cool tutorial on the FOR instruction. This instruction is taken out of a very common language, such as C++ and Java. However, in ladder logic, it could be a little bit confusing because the structure is a little bit different. However, once you've implemented it once or twice, you will know exactly what it is, its purpose, and essentially how to apply it in order to make your programs a lot more efficient.

FOR is an advanced instruction. It is not something I would recommend you would be using on a regular basis. But it is something that you need to be familiar with because it can greatly simplify your coding. Now it's time to get right into it:

Video 42: PLC Programming Advanced Topics - FOR - FOR Loop Development

4.2. PLC Forces - What you Need to Know

In this section, we're going to be talking about PLC Forces. Forces is an extremely important topic and is present on most PLCs. Fig. 4.2.1 is my CompactLogix L24ER which I have on my desk right now.

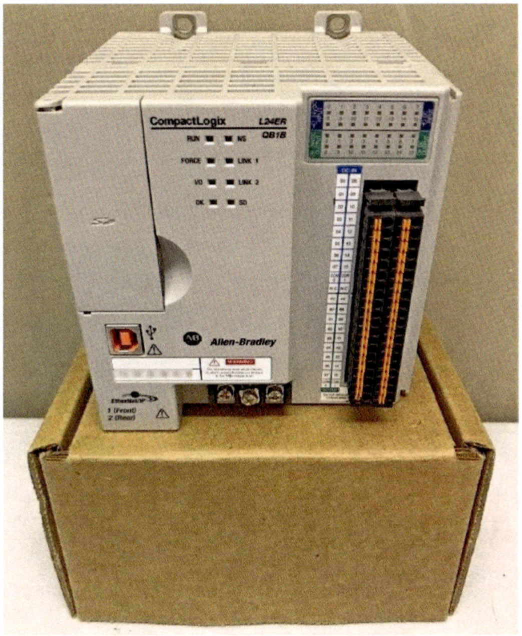

Fig. 4.2.1: CompactLogix L24ER

I'm going to be live with this PLC, but we're going to be discussing what forces are first. Then we're going to jump into a couple of demonstrations. Let's get started now:

Video 43: PLC Forces - What you Need to Know

4.3. Implementing Add-On Instructions in Studio 5000

Welcome to another advanced PLC programming tutorial in which we're going to focus on **Add-On Instructions**, which are more commonly known as the AOIs. From your RSLogix 5000 or Studio 5000 environment, you will find the Add-On Instructions folder on the left-hand side within the pane.

Watch this tutorial video to understand the Add-On Instructions well:

Video 44: Implementing Add-On Instructions in Studio 5000

4.4. Input/output Buffering: Best Practices for PLC Inputs & Outputs

This section is about input and output buffering. You should have seen the implementation of this input and output map in my programs that I've done previously. You've probably encountered this practice in the online programs that you've seen, as well as probably some of the stuff that you've seen at work.

In programming PLCs, we need to discuss why we do want to buffer our inputs and outputs, as well as some of the key advantages that I've personally had to use and have seen in practical experience.

We're also going to look at a PLC in order to fully understand this concept. So, in order to understand the software, we're going to briefly look again at the hardware in Fig. 4.2.1. It is the same PLC that I've been programming for quite a few times in my previous tutorials.

Here's the detail:

Video 45: IO Buffering best Practices for PLC Inputs and Outputs

Chapter 5. Large PLC Programming Project - PLC-HMI Street Traffic Light System

The whole of this chapter is devoted to the development of a very large project. As you probably have seen from the chapter heading, it is a traffic light system. In other words, we're going to develop a control system which is going to automate a street traffic light system.

It is very simple to understand. Therefore, this it makes it a very cool system to work with. That way, I don't need to explain all the details and intricacies of what I'm doing.

There's a link posted in Chapter 6 which you can use to get the latest files of this and other projects so that you can be up to speed with the tutorials, and so that you can go back and understand what was done.

5.1. Initial Setup for Green, Yellow & Red Lights Logic - Part 1

Here's a couple of different things you need to know before we begin. We are using Studio 5000 Logix Designer. We're also running this project live on a 1769-L24ER- QB1B processor, version 30.11. Of course, you can use any other type of processor. This project not going to be specific to Allen-Bradley. So, you can replicate this same project on any other processor of your choice.

That being said, let's start go to the tutorial now and start working on the technical setup:

Video 46: PLC HMI Program Project Init Setup for Green, Yellow and Red Lights

5.2. Hand-Shaking Logic & HMI Interface Tags - Part 2

This is the second tutorial on our project on a street traffic light system, in which we're programming a PLC to control a traffic light intersection. In the previous video, just to quickly recap, we created two different lights, which essentially control the traffic from South to North, and the traffic from East to West respectively.

These two lights are essentially your standard traffic lights, which turned from green to yellow, followed by a red. In this section, we're going to be implementing a system which essentially handshakes between the two lights. Without further delay, let dive in:

Video 47: PLC HMI Programming Project - Hand-Shaking Logic and HMI Interface Tags P2

5.3. Initial HMI Screen Development & Animations - Part 3

In the previous section, we built a simple HMI interface on our CompactLogix PLC. As you will see in the following video, it cycles through the three different colors on two individual lights, Light1 and Light2.

In this section, we're going to interface this back to our HMI system. I'm going to show you how to link the HMI back to the PLC and be able to display some of these tags. We're also going to work with different images and different tags, so open the next video so you can learn quite a bit when it comes to FactoryTalk View development.

Video 48: PLC HMI Programming Project - Initial HMI Screen Development and Animation P3

5.4. HMI Screen Layout & Design - Part 4

In the last video, we animated one of our first lights based on the API that we've created on the PLC side. We animated all of that in FactoryTalk View studio. In the next video you will see it playing. It's a very simplistic animation, but it does allow us to essentially link the HMI to the PLC.

Now we're going to look at adding the rest of the elements, and be able to create a simple animation, which is going to be based on our PLC system. Without any further delay, let's get right into this video:

Video 49: PLC HMI Programming Project - HMI Screen Layout and Design Part 4

5.5. HMI Control System Development - Part 5

In this the project tutorial part 5, we're going to continue to work on our light system. As a quick reminder, what we currently have is the PLC program that is switching between two distinct lights: one which is going from North to South, and the other one which is going from East to West.

In this section, we're going to implement a control scheme. We currently have only an overview page which allows us to see the status of the lights, but we cannot modify any of the parameters on our system. So, we're going to focus on allowing some kind of a control which will be sending signals from the HMI system back to the PLC and allowing us to change certain parameters.

It's often seen as a good practice to control the HMI from a different screen, essentially a popup, rather than from the main overview screen. There's a couple of reasons for that. First of all, you might want to enable certain security features, such as user login or log out, and you want certain users not to be able to make changes.

However, in this project we're just going to implement the buttons; we're not going to focus on security as much. Also, as said before, the control screen will be separated so that you can't inadvertently click on them without going to that specific screen.

Let’s get right into it:

Video 50: PLC HMI Programming Project - HMI Control System Develop Part 5

5.6. Control System Display Development, Numeric Entries & Displays - Part 6

We are now on part 6 of our project development, in which we're talking about PLC and HMI development for a street light traffic system. In the previous video, just as a quick reminder, we've created a secondary screen from which we're going to control certain parameters on the PLC.

A code is going to be written by the HMI to the PLC in order to change the values of our timers. We're going to be working on the screen. We will be writing certain tags back to the PLC. Let's start right now:

Video 51. PLC HMI Program Project - Control System Display Development, Numeric Entries and Displays Part 6

5.7. Adding Physical Inputs & Outputs for the System Programming Tags to Handle Local I/O - Part 7

Welcome to the episode 7 of our PLC and HMI development project, in which we're building a simple street light traffic system. On the PLC side, just as a little reminder, we have code, which is alternating between green, yellow, and red on two different lights.

On the HMI, we have a normal street light traffic system which goes to yellow and to red, while the other one changes to green, yellow, red, and then they alternate one after another. Let's continue the project in the next tutorial video:

Video 52: Adding Physical Inputs & Outputs for the System Programming Tags – Part 7

5.8. System Fault Handling Implementation - Part 8

In the previous episode, we've inputted two different cards which are responsible for the inputs, as well as the outputs for controlling our system. That being said, the PLC definitely does faults out if you don't have the hardware installed because that fails to communicate. So, I had to change to local PLC tags.

Let's finish this project up in this tutorial:

Video 53: System Fault Handling Implementation - Part 8

Chapter 6. How to Download the Training Videos, PLC Programs (Codes) & Demo Editions of RSLogix 5000 / Studio 5000 Logix Designer

Below are the links to download all the 53 training videos for this section, PLC programs and the Demo edition of RSLogix 5000 / Studio 5000 Logix Designer:

https://bit.ly/3CZ5frw – When you open this link, you should be able to download:

1. All the 53 training videos/tutorials (in mp4 format)

2. Copyright and License Info (txt file)

3. PLC Program files (codes), titled "My_PLC_Development_Zipped" (Zipped file)

Here's the link to download any of the **demo versions** of Logix designers and emulators:

Download Various Demo Versions of RSLogix 5000 and Emulator Designer or use this short link: http://bit.ly/2lTyw3F.

I use version 20 and version 17 in most of my projects. Version 17 has an expiry time of 90 days. It also has a few limitations, but you will be gaining experience with the real thing. Here is the link to download it:

Click here to download RSLogix 5000 Designer V17 or use this short link: http://bit.ly/2lyjosq. During the installation of Version 17, you will be asked for **serial number**. Just enter a random number such as 123 or 12345. You can download the full installation guide here or use this short link: https://bit.ly/35VHVg6.

Chapter 7. How to Get Further Help

Dear reader, I hope you find this section and the accompanying videos very helpful for learning PLC programming using Allen-Bradley's RSLogix software and hardware. You may find some of the concepts I covered in this section confusing at first. However, with time and a little more effort, you should be able to program a PLC to do whatever you want it to do.

Contact me through my email below or visit my Author page if you need further help regarding the topics covered in this section, or if have questions or requests.

Regards,

amandamorin3000@gmail.com

Charles J. J.

Section 2: How to Setup, Integrate & Program the Most Used Allen Bradley PowerFlex 525 Drive

This section is an exhaustive collection of tutorials and training videos on how you can advance your knowledge of PLCs by working with PowerFlex 525 family of Variable Frequency Drives.

There are 5 chapters in this section. They are accompanied with **16 in-depth HD training videos that you can download**. These videos simplify everything you need to understand, and help you **speed up your learning** of Allen-Bradley's PowerFlex 525 drives and how to install them within a manufacturing environment. There is also a link at the end of this section for you to download my PLC programs (codes) for your revision.

Since I assume you have little knowledge of PowerFlex 525 Drive and PLC programming, I prepared this section in such a way that when you read it and study the accompanying training videos (16 episodes), you will not only have an in-depth knowledge of the different parameters which need to be configured in order to properly setup and utilize the PowerFlex 525 VFD, you will be able to make sense of the documentation, and gain a lot of job experience you need to build innovations and earn higher salaries.

In this section, I start with the basics, that is, connecting power and turning on the PowerFlex 525 hardware, and then move on to the control methods that don't even require you have the hardware. Then I proceed to the advanced control methods that utilize the EtherNet/IP protocol, as well as a CompactLogix 1769-L24ER-QB1B PLC. This will help you develop confidence in working with these Variable Frequency Drives.

So, after studying this section and the training videos, you should develop a hands-on approach to wire a PowerFlex 525 drive, setup the parameters and control it with a Programmable Logic Controller (PLC) and be able to use it in real world industrial applications.

The methods presented in the training videos are those that are usually employed in the real world of industrial automation, and they may be all that you will ever need to learn. The information in this section and the training videos is very valuable to any skillful PLC programmer, no matter their skill level.

However, if you are a complete novice to PLCs or PLC programming, here's the link to my book that you will find very helpful to get started: PLC Programming Using RSLogix 500 & Industrial Applications: A Guide for Learning Ladder Logic Concepts Step by Step . Then you can move on to the first section of this book.

You will find the link to download all the 16 training videos and program codes in Chapter 4 of this section. The serial number and title of the video you should watch is mentioned in the appropriate section of each chapter.

For effective learning and to get good practice, I strongly advise you attempt the challenges and project in this book on your own. Build your own program after you have watched me build mine in each video. Then later you can cross check your programs with mine. Also feel free to use or modify any of the PLC programs you downloaded as you wish.

Finally, if you have questions or need further help, use the support link I provided at the end of this section. I will get back to you very quickly.

Chapter 1. Hardware Overview & Getting Started

First, we need to turn on the PowerFlex 525 drive before we can do anything with it. Fig. 1.1 shows the Powerflex 525 VFD (Variable frequency Drive).

Fig 1.1: Allen-Bradley PowerFlex 525 VSD

This PowerFlex 525 AC Drive has a very innovative and modular design. It can be easily and quickly configured and installed. It is a new generation of compact drives that offer embedded EtherNet/IP communications. It can be programed via USB ports, **and also** has standard safety features. All PowerFlex 525 drives have power rating between 0.4 and 22 kW (between 0.5 and 30 Horsepower).

It can be powered with global voltage, in the range of 100 to 600 Volts. One of its main uses is motor control, and it has flexible mounting options. All PowerFlex 525 AC drives can be operated at up to 70 °C (which is 158 °F) **temperature**. They are therefore ideal for applications like fans, mixers, conveyors, pumps, and lots more.

1.1. PowerFlex 525 - Connecting Power & Turning On the VFD

To turn a PowerFlex 525 on, we need to send some power. In the first demo video below, we bring a standard 110 volts AC power into the drive. Let's get started:

Video 1: PowerFlex 525 - Connecting Power and Turning On the VFD – 4:18 Minutes

1.2. PowerFlex 525 Hardware Overview

In the next PowerFlex 525 hardware overview video, we are going to look at the hardware mor closely, and study how it plays into communicating with a motor, as well as the PLC. We will go to go into a lot of detail and learn what everything is.

This PowerFlex 525 variable frequency drive is a single-phase input, and 3-phase output drive. This makes it easier for home development because it can be plugged into 110 volts AC, so we are able to program our power flex drive at home.

Of course, in a real industrial world scenario, it is more common to find a 3-phase input and a 3-phase output, but they look identical from the surface. Without further delay, let's see the demo video now:

Video 2: PowerFlex 525 Hardware Overview – 6:40 Minutes

1.3. PowerFlex 525 - Wiring a 3 Phase Motor to the Variable Frequency Drive

Now that our drive is powered on, I believe it's always a lot more interesting to look at a drive that actually powers a motor. So, in the next video, we're going to talk a little bit about an *Oriental Motor* (shown in the video) which is made in Japan.

We're going to look at its label first, and then how it is wired inside. It is an induction motor, as you might expect, and it is also a 3-phase input. Also, it can be empowered by different voltages. Fig. 1.3.1 shown how the OM motor is wired to the PowerFlex 525 VFD.

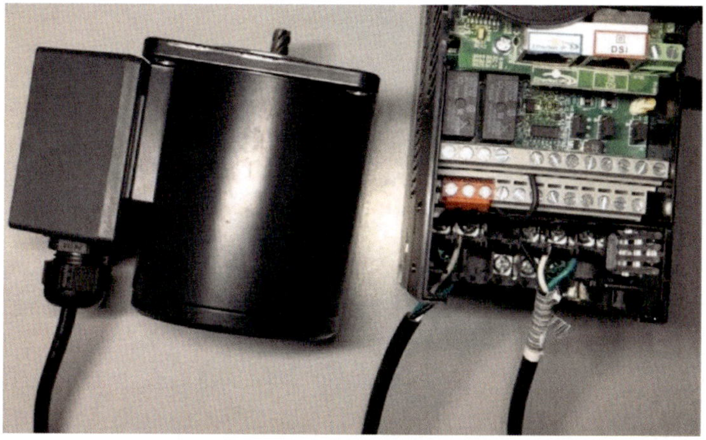

Fig. 1.3.1: The 3 Phase OM Motor (left) is wired to the PowerFlex 525 Variable Frequency Drive (right)

Now let's jump right in to learn the detail:

Video 3: PowerFlex 525 - Wiring a 3 Phase Motor to the Variable Frequency Drive – 5:54 Minutes

1.4. PowerFlex 525 Quick Start Documentation Walkthrough

In this section we're going to take a step back and look at a quick start manual for PowerFlex VFD drives. In this manual, you will find everything you need in order to quickly get started with your motor. In this documentation, there is a lot of parameters that you can remember and memorize.

In practice, I always refer back to this manual whenever I start a new motor in order to make sure that everything is okay in the perspective of the PowerFlex. However, we're not going to go into a lot of depth, but we are going to look at a few key components of this manual.

So, you can definitely look it up on Google, or you can just use the link in Chapter 4 to download this manual. Of course, there's going to be a lot of general precautions, such as not to work with live drives, not to start disconnecting live drives and so on, and that a VFD drive is something that should be only attempted by qualified individuals.

Now that we've taken the basic introduction out of the way, let's go to the demo:

Video 4: PowerFlex 525 Quick Start Documentation Walkthrough – 9:39 Minutes

1.5. PowerFlex 525 Basic Parameter Setting for Motor

In this section, we're going to look through the basic parameters that I mentioned in the last demo. This is something that you want to go through for every single motor, but as you get more and more experienced, you will notice that a lot of these remain the same from the factory setup. So, you can leave them as such.

In the next demo, I'm going to power on my motor with the switch on the right side of the VFD. Watch the demo now:

Video 5: PowerFlex 525 Basic Parameter Setting for Motor – 9:39 Minutes

1.6. Starting & Stopping the Drive through Digital Outputs of the PLC

In the next demo, we're going to explore one of the simplest ways of controlling our PowerFlex 525 drive from an external source. Although it is highly recommended to use ethernet to control these drives, there is still a lot of facilities which prefer the old way: a digital on and off method.

There's nothing wrong with that if you're using a certain motor that runs a constant speed level, you don't necessarily need to get into the full ethernet-based control. Also, perhaps you don't have a device which can do ethernet IP communications. So, you want to simply turn the drive on and off, and be able to turn on a pump, which is going to be running at a constant speed. Now let's open the next demo and keep things moving:

Video 6: Starting and Stopping the Drive through Digital Outputs of the PLC

1.7. Running the Drive in Reverse through a Digital Output

In this section, we're going to talk about how to use a digital output, just like we did in the previous demo, but to reverse the drive direction. So, as you've probably noticed, when we try to start and stop the drive, we always have a forward connotation (icon) on the lower left corner of the display of the PowerFlex drive. This essentially means that the drive is only running in one single direction.

Let's look at how to run the drive in reverse quickly in the next demo:

Video 7: Running the Drive in Reverse through a Digital Output

1.8. Setting a Speed Reference from the Keypad instead of Potentiometer

Now we will look at the basic speed referencing for PowerFlex 525 drives. In the last video, we were able to start and stop this drive. Although we were unable to fully control the speed of the drive, we were able to set it up through the potentiometer.

This is reflected by the **[Speed Reference1]** parameter (**P047**) as you will see in this video:

Video 8: Setting a Speed Reference from the Keypad instead of Potentiometer

Chapter 2. Variable Frequency Drive (VFD) Control from a PLC Over EtherNetIP

In the last set of demos, we were able to start and stop, as well as control to a certain extent, the speed of our drive through a digital set of inputs and outputs.

2.1. EtherNet_IP and Other Methods of Control Introduction

The preferred method that you want to use with these specific drives is definitely going to be EthernetIP, which is listed under number 15 in the speed reference section.

EthernetIP can be used for a whole array of things. Therefore, I have dedicated a whole section for it which we're going to explore in a lot of depth. Without further delay, let's go over to the demo to learn this EtherNet_IP and other methods of control:

Video 9: EtherNet_IP and Other Methods of Control Introduction

2.2. Establishing an EtherNet_IP Connection to the PowerFlex 525 Drive

This section is a continuation of our study on EthernetIP. Now, we're going to establish a network connection to our drive. First and foremost, we're going to power it on and see what kind of settings we have. In the video, you will notice that the drive is going to go through its boot sequence as usual.

In order for us to start communicating over EthernetIP however, we need to give the drive a specific IP address, as well as a subnet mask. Let's go over to learn the details in this demo:

Video 10: Establishing an EtherNet_IP Connection to PowerFlex 525 Drive

2.3. Verifying Communication, Setting Parameters & Visualizing RSLinx Communication

This is another section on networking our PowerFlex 525 drives. In this section, we're going to explore how to communicate between our computer and the PowerFlex 525, and how to be able to see it through RSLinx, as well as some of the other PLC tools.

In the previous demo video, we talked about how to set the parameters of this specific drive. We set them to an address, which we are deeming essentially usable on the private side network. But now we need to be able to communicate, send some parameters, and check that the driver is there.

This is actually a best practice and something that I do in the field every single time. This is because I want to make sure that the IP address matches the drive that I expect in the field. So, every single time that I set up these drives, I go through this exact same procedure in order to double-check everything.

Now in order to get started, let's dive in to the next demo:

Video 11. Verifying Communication, Setting Parameters & Visualizing RSLinx Communication

2.4. Adding the PowerFlex 525 Drive to the Studio 5000 Project and Going Online

In this section, we're going to add our drive to the Studio 5000 PLC program, which you will see in the next demo video in just a moment. But first, a little reminder. We have a PowerFlex 525 drive, which is connected through an ethernet cable to our CompactLogix L24ER-QB1B PLC.

This PLC is running version 30.11 of Studio 5000 software. So, we're going to be communicating from this PLC to our PowerFlex 525 drive. Without further delay, let's jump right in:

Video 12: Adding the PowerFlex 525 Drive to the Studio 5000 Project and Going Online

2.5. Configuring Drive Parameters, Starting, Stopping & Using a Speed Reference

In this section, we're going to start to control our drive through EthernetIP, through the use of our PLC, which runs Studio 5000 version 30.11. If you remember, a couple of demos ago, we used the front panel of our drive in order to make some parameter changes. We gave it an IP address. We also set a start command source.

In the next demo video, we're going to be changing all that, in order to be able to communicate and control the drive over ethernet. Let's look at it right away:

Video 13: Configuring Drive Parameters, Starting, Stopping & Using a Speed Reference

Chapter 3. Programming PLC Control for the PowerFlex 525 VFD Studio RSLogix 5000

In this chapter, we will look at programming PLC control for our PowerFlex 525 VFD in the Studio RSLogix 5000.

3.1. Flashing the Firmware of the VFD 1.003 -- 5.002 - ControlFlash Software

Now, we're going to look at flashing the firmware of our PowerFlex 525. There is a reason behind flashing the firmware. For example, to get access to certain different features and to be able to use different parameters which are not available in the earlier revisions.

Furthermore, you will also be able to avoid any problems that come with the earlier versions, which have been upgraded for a number of reasons. This process actually doesn't only apply to the PowerFlex 525 alone, but also to many other Rockwell automation, tools and hardware.

Since you will learn quite a bit from this demo, let's just jump into it:

Video 14: Flashing the Firmware of the VFD 1.003 -- 5.002 - ControlFlash Software

3.2. Basic Ladder Logic Implementation of VFD Control - ControlFlash Software

In this advanced section, we're going to start programming our drive in ladder logic. This is a section in which we're going to be fully going over the process and good practices of how we can control our PowerFlex.

We've looked at some basics and we've toggled some bits, but ultimately, we want to be able to write some logic, which is going to give a user interface and give the operator, or whoever is working on the manufacturing floor, the ability to start and stop the motor in manual and auto mode.

Of course, in auto mode the motor's going to be started and stopped by the program. I'm going to assume that you are not extremely familiar with ladder logic, although I'm sure that you are fairly experienced.

Now I have opened my Logix Designer Studio 5000 on a PLC, which is running version 30.11. Let's go to the next demo now to learn the program:

Video 15: Basic Ladder Logic Implementation of VFD Control - ControlFlash Software

3.3. PowerFlex 525 VFD Fault Handling and Status Logic - ControlFlash Software

In the previous section, we were able to program some very simple ladder logic in order to be able to start, stop and write a speed reference to our drive. But that's not all you have to do with these drives. A lot of times there will be a point at which one of these drives will fail, and you do want to have the ability to know when it has faulted. So, there's going to be a couple of things that we need to implement. First of all, we want to implement a status for this specific drive.

Without further delay, let's jump in to the next demo to learn more:

Video 16: PowerFlex 525 VFD Fault Handling and Status Logic - ControlFlash Software

Chapter 4. How to Download the Training Videos, PLC Programs (Codes) & Demo Editions of RSLogix 5000 / Studio 5000 Logix Designer

Below are the links to download all the 16 demo videos for this section, PLC programs and the Demo edition of RSLogix 5000 / Studio 5000 Logix Designer:

https://bit.ly/3j2c2ZD – When you open this BIT.LY link, you should be able to download:

1. All the 16 demo videos/tutorials (in mp4 format)

2. Copyright and License Info (txt file)

3. PLC Program files (codes), titled "My_PLC_Development_Zipped" (Zipped file)

4. PowerFlex 525 User manual (pdf file)

Here's the link to download any of the **demo versions** of Logix designers and emulators:

Download Various Demo Versions of RSLogix 5000 and Emulator Designer using this BIT.LY short link: http://bit.ly/2lTyw3F.

I use version 20 and version 17 in most of my projects. Version 17 has an expiry time of 90 days. It also has a few limitations, but you will be gaining experience with the real thing. Here is the link to download it:

Use this BIT.LY link to download RSLogix 5000 Designer V17: http://bit.ly/2lyjosq.

During the installation of Version 17, you will be asked for **serial number**. Just enter a random number such as 123 or 12345. You can use the following BIT.LY link to download the full installation guide: https://bit.ly/35VHVg6.

Chapter 5. How to Get Further Help

Dear reader, I hope you find this section and the accompanying videos very helpful in learning how to setup, integrate and program the most commonly used Allen Bradley PowerFlex 525 Drive with my training Videos. You may find some of the concepts I covered in this section confusing at first. However, with time and a little more effort, you should be able to program a PLC to do whatever you want it to do.

Contact me through my support email below or visit my Author page if you need further help, or if have questions or requests.

Regards,

Charles J. J.

amandamorin3000@gmail.com.

Section 3: How to Develop & Embed Machine Vision System in PLC with Training Videos

This section is an exhaustive collection of my tutorials and training videos on how you can develop and embed the Cognex In-Sight Machine Vision System in Programmable Logic Controller (PLC).

There are 6 chapters in this section. They are accompanied with **23 in-depth HD training videos that you can download**. These videos simplify everything you need to understand, and help you **speed up your learning** of Cognex In-Sight Machine Vision for industrial development. There is also a link in this book for you to download my PLC programs (codes) for your revision.

I assume you have little knowledge of Machine Vision application to PLCs. So I prepared this book in such a way that when you read it and study the accompanying training videos (23 episodes), you will not only have an in-depth knowledge of the different parameters which need to be configured in order to properly connect and communicate a Cognex camera to your PLC, you will also learn how to purchase, upgrade/downgrade the device firmware and trigger the camera. This will help gain a lot of job experience you need to build innovations and earn higher salaries.

I start with the basics, that is, an overview of the Cognex In-Sight 7000 Series Camera, and then move on to the detail of the In-Sight Software. Then I proceed to the Cognex pattern and part inspection tools, as well as how to trigger the Cognex In-Sight Camera. I went as far as dedicating a whole chapter to a 3-part in-depth tutorial on how to read bar codes with the camera.

So, after studying this section and the training videos, you should develop a hands-on approach to the installation and integration of Cognex into Studio 5000 Rockwell Environment, and be able to use it in real world industrial applications.

The best way to master PLC programming is to use real world situations. The real-world scenarios and industrial applications developed in this book and its accompanying video trainings will help you learn better and faster many of the functions and features of Studio 5000 platform.

The methods presented in the training videos are those that are usually employed in the real world of industrial automation, and they may be all that you will ever need to learn. The information in this book and the training videos is very valuable, not only to those who are just starting out, but also to any other skillful PLC programmer, no matter their skill level.

Merely having the user manual of Cognex In-Sight 7000 Series Camera, or referring to its help contents, is far from enough in learning machine vision development. Therefore, this book and the accompanying trainings are extremely useful for learning machine vision development and its applications.

However, if you are a complete novice to PLCs or PLC programming, here's the link to my book that you will find very helpful to get started: PLC Programming Using RSLogix 500 & Industrial Applications: A Guide for Learning Ladder Logic Concepts Step by Step . Then you can move on to the first section of this book.

You will find the link to download all the 23 training videos and program codes in Chapter 5 of this section. The serial number and title of the video you should watch is mentioned in the appropriate section of each chapter.

For effective learning and to get good practice, I strongly advise you attempt the challenges and projects in this book on your own. Build your own program after you have watched me build mine in each video. Then later you can cross check your programs with mine. Also feel free to use or modify any of the PLC programs you downloaded as you wish.

Finally, if you have questions or need further help, use the support link I provided in Chapter 5 of this section or elsewhere in this book. I will get back to you very quickly.

Chapter 1. Hardware Overview: Getting Started with Cognex In-Sight Camera

In this chapter we're going to look closely at the Cognex In-Sight 7000 Series Camera shown in Fig. 1.1.

Fig. 1.1: Cognex In-Sight 7000 Series Camera. Cognex.com

We will talk about the pin outs on the camera, as well as its capabilities. But before we jump into the hardware, I believe it's very important to answer this question:

What are the purposes of a vision system, especially within an industrial or manufacturing environment where they are most commonly deployed?

To answer that question, it's very important to look at the manufacturers themselves and see what they recommend or say about their systems. If you take a look at the Cognex website (http://bit.ly/3juNdUK), you will see different products, in particular the In-Sight product line. There are quite a few different systems available for your selection.

1.1. First Look at the Cognex In-Sight Camera Series

On the Cognex site, a description of the features of the In-Sight 7000 Series says it's a powerful, modular vision system that is highly field-customizable to your application requirements. But that doesn't really answer the question as to what the system does. However, from industry experience, I can tell you that there are 3 main things that your system is going to be capable of doing.

The first one is to **detect** what's called a reject or quality defect. That is, a product that does not correspond to the specifications to which it was meant to be built. Here are some examples.

Imagine a bottle which gets filled to a certain level where the manufacturer has a certain specification. Take a bottle of one liter of Coke as an example. It is normally filled at one liter but is allowed to be slightly below or slightly above that level. You don't necessarily always hit the target, but you do have a certain limit range of the bottle. So, a vision system can be placed to look at the bottle which is moving on a conveyor. If the system detects that the bottle is either below or above a certain threshold, then it is rejected due to a quality defect.

Of course, that is just a very basic example. I've seen some parts, such as precisely manufactured bolts for example, being inspected to a certain tolerance. So, if we have a bolt that needs to be down to, let's say, 0.001mm dimensions for aerospace purposes, then a camera will be placed to inspect if those tolerances are met. That's a quality or a defect examination.

The second thing that you can do with cameras is **measurements**. For example, you can measure things like the diameter or length of a certain part. Once again, that is commonly used for rejects, but it can also send information back to your system to help you analyze a statistical data.

For example, if you're making plastic spoons and you start deviating from the form factor of that spoon, you can predict the kind of an adjustment you need to make on your plastic forming machine in order to correct that particular deficiency and get that system back into check. This is not per se a quality check but a *measurement* check that you can do on different kinds of your parts.

The third application is going to be **machine specific**. For example, I had an application where I had a certain rapper going around a particular piece. That wrapper needed to be positioned in a certain place. So, I needed to take a measurement of where that deviation is, and essentially correct the web in order to wrap that piece in the right position. This is essentially used for control, not rejecting the product, because we are essentially controlling a piece of equipment.

Another example of such an application would be in **robotics**. Picture a conveyor with a robot at the end. On this conveyor, you have parts that are scattered in different orientations. Your vision system can look down and examine where the pieces are relative to the speed of course of the conveyor. It can tell the robotic arm to, for example, pick up a piece and place it in a particular orientation, and so on and so forth. It's essentially providing feedback for the control system to work efficiently.

Those are the three main applications that I've seen in most vision systems. But I do encourage you to explore the Cognex website. There're different data sheets you can download. There're different specifications of this camera that you can look at.

Of course, Cognex does describe the full capabilities, which I sometimes look at when I do need to pick a new system. The camera that I own is a monochrome (black and white), but you can of course get vision systems with color. You can get, depending on your needs, a whole, a wide array. Of course, there's always trade-offs, depending for example on how fast you want to process your part, and how complex your application is.

So, I highly encourage you to explore this particular line, as well as the other lines from Cognex, in order to see which one fits your needs the most.

1.2. Overview of the Cognex In-Sight 7000 Series Camera

This section is the full overview of the camera itself, which we are going to use throughout this book. We're also going to talk about the different variations of these cameras. I'm going to talk about the connections that you need to know how to make in order to make this camera function. So, let's jump right into the first demo video for the full overview.

Video 1: Overview of the Cognex In-Sight 7000 Series Camera

1.3. Detailed Look at the Hardware Connections & Pinout from the Cognex 7000 Camera

In this section we will look at the power and input and output breakout cable, which connects to the bottom of the Cognex camera that we saw in the previous demo. It is a 12-pin cable, which allows you to send quite a bit of information back from your camera. As the name would suggest, it powers the camera. In this demo, we will look at the different wires we have in this bundle.

Video 2: Detailed Look at the Hardware Connections & Pinout from the Cognex 7000 Camera

1.4. Connecting the Cognex Camera to a 24VDC Power Supply

In this section, we're going to finalize our connections to the camera in order to be able to power it on, and of course, go online with it from our computer. We will use a 1606-XLP 24-volt DC power supply (with 15 watts of power) from Allen Bradley.

In this demo, I want you to turn your attention to my desk because I'm going to walk you step-by-step on exactly what you need to do in order to get your device up and running. Let’s open the demo now so I can show you all the connections you need to make, which we've explored in the previous video.

Video 3: Connecting the Cognex Camera to a 24VDC Power Supply

1.5. A Guide to Purchasing a Cognex In-Sight Series Camera

This short guide will help you with the purchase of a Cognex 5000 Series In-Sight Camera system by outlining key components, pointing out the different part numbers and defining what's recommended as a practice kit.

System Hardware

You will need 4 main components to get started with your Cognex In-Sight system:

1. Cognex Camera

2. Cognex Lens

3. Cognex M12 EtherNet cable

4. Cognex M12 Breakout cable

5. 24VDC Power Supply

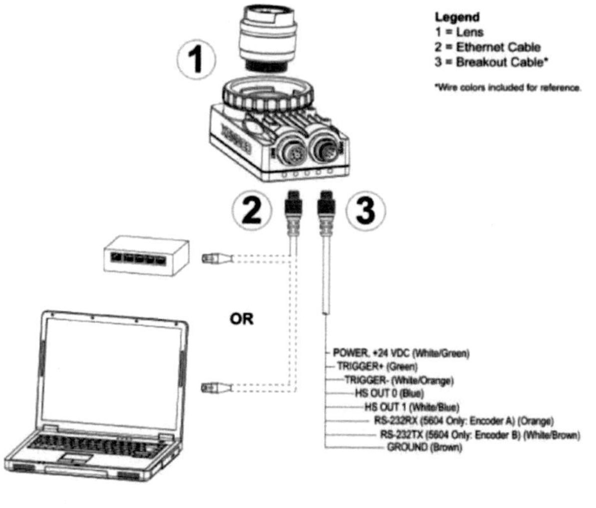

Fig. 1.5.1: The 4 main components to get started with your Cognex In-Sight system

Cognex Camera

At this time, Cognex is offering multiple lines of 2D vision cameras: 2000 series, 5000 series, 7000 series, 8000 series, 9000 series, etc. However, if you're looking to buy a camera for personal use on eBay, you'll notice that the spectrum is much wider. You may find 1000 series cameras and even DVTs.

Although any of these cameras may be an option, depending on your budget, I'd highly recommend the 5000 series camera. It's inexpensive, has a robust profile and is generally more capable than the micro series (1000).

Key Differences Between 5000 Series Models

1. **Resolution**: You will find that cameras of 5100, 5110, 5400C, 5400, 5410, 5600 and 5610 series will have a lower resolution of 640 x 480. The 5403, 5413, 5603 and the 5613 cameras will have a resolution of 1600 x 1200. Lastly, the 5605 and the 5615 camera models will have a resolution of 2448 x 2048. As you can probably guess, you will be charged a premium for resolution. Purchasing a low-resolution camera for practice purposes is absolutely fine.

Note: Higher resolution cameras will typically be monochrome (See below).

2. **Color**: Cameras will come in either a black/white (monochrome) configuration or a full color one. The part number will have a trailing C which will indicate if the camera is full color or not. "C" = color.

Note: A color camera will typically have a lower resolution since the pixel density needs to be smaller. (See above).

3. **Frames Per Second**: Cameras will typically have 60 full frames per second. However, certain models, typically the ones with a higher resolution, will only provide 14 FPS. Using either one is fine.

Cognex Lens

Fig. 1.5.2: A Cognex Lens

Cognex doesn't manufacture lenses and the ones they sell are typically on the expensive side. The lens is extremely important and it's important to get this right. You will have three parameters at your disposal when it comes to static lenses: working distance, field of view height and field of view width. In other words, how far will the object be and how big will it be.

Choosing the right lens will depend on what you're trying to accomplish; I'd highly recommend that you look at the Cognex selection tool once you've decided on a lens and make sure that the above parameters are correct.

Note: Cognex does provide auxiliary tools for lenses, such as extension kits. These may be used to extend the range of a lens. Similarly, you may also want to purchase an LED or light ring which is optional.

Link to tool:

https://www.cognex.com/resources/interactive-tools/lens-advisor.

Cognex M12 EtherNet Cable

Fig. 1.5.3: A Cognex M12 EtherNet Cable

You will need a cable which goes from the Cognex Camera into an RJ45 EtherNet port. This could be your laptop or a switch. Cognex has variable lengths of this cable; either one is fine. The original Cognex cable will be of an iconic green; I'd recommend purchasing the branded cable.

Cognex M12 Breakout Cable

Fig. 1.5.4: A Cognex M12 Breakout Cable

To power the camera, you will need the breakout cable. It's also used for IO points as well as the trigger. The cable comes in various lengths and should have unterminated wires at the end. I'd recommend buying the original Cognex cable.

24VDC Power Supply

You will need an external power supply for the camera. Cognex doesn't manufacture their own, so any power supply manufacturer should be OK. It's recommended to get a power supply with an output of at least 1A as the Cognex can consume up to 500mA and you may want to use the remainder of the current for a sensor, trigger, push button, etc.

Important Notes

1. All the Cognex software and firmware is available for free on their website, don't purchase anything of this kind.

2. Don't purchase a DVT camera. This is an older model which was purchased by Cognex and the configuration of these cameras is completely different. They are often labeled as Cognex on eBay and many make the mistake of purchasing these instead of the In-Sight versions.

3. You don't need the IO breakout module. It allows one to expand the breakout cable to multiple inputs & outputs. It may be applicable to certain field application but is completely unnecessary for most cases based on my experience.

Chapter 2. In-Sight Software & Getting Started

In this chapter we're going to talk about how to download the In-Sight software, which corresponds to the Cognex In-Sight 7000 Series Camera.

2.1. In-Sight Explorer Download Instructions

In the demo, I'm going to show you briefly where to get the software. To start, go to www.cognex.com. Locate and click the *Products* link on the top menu. Then scroll down to see the different vision systems. Please go to the demo now to watch the step by step instructions.

Video 4: In-Sight Explorer Download Instructions

2.2. Establishing an EtherNet Connection to the Camera

What we're going to do in this section is establish an ethernet connection to our camera. This is one of the most difficult steps sometimes when dealing with different hardware.

So, in the demo I explain the process clearly to make sure that you understand this step very well. I use the insight Explorer that we installed in the previous demo. Let's begin now:

Video 5: Establishing an EtherNet Connection to the Camera

2.3. Upgrading or Downgrading the Firmware on the Device

In this section we will be going over a critical step in order to get our hardware communicating the way we want it to be. In the last episode, we went online with the camera and made sure that everything was connecting as we would expect.

That being said, there is often a need to update the firmware. This is because sometimes you buy a sensor off eBay, or maybe you just get it brand new from Cognex, and it's just not at the same firmware level as you would expect, or it just doesn't communicate well with your software package.

Therefore, it is always important to first make sure that the firmware level you're running is compatible with the software you're running on your computer. It is also important to make sure that you're running the latest firmware. So, let's go over to the demo and see how all this is done:

Video 6: Upgrading or Downgrading the Firmware on the Device

2.4. Setting Up the Image Part 1

Now that we are back online with the camera after upgrading the firmware in the previous demo. All I had to do was click this connect button in order to establish a communication with the camera. Without any further delay, let's go over to the demo to see how to set up the image.

Video 7: Setting Up the Image Part 1

2.5. Setting Up the Image Part 2

In this section we are going to continue setting up the image for our parts. We will use a small screwdriver to demonstrate the setup feature of the camera. Let's dive right in:

Video 8: Setting Up the Image Part 2

2.6. Locating the Model (Part) through Cognex Pattern Tool

Welcome back to this Cognex camera training. In the last demo, we actually finalized the "Set Up Image" button. I fine-tuned it just a little bit so we could get the proper contrast on our tool.

But in this section's demo, we are going to move in to the "Locate Part" tool. Let's go to the demo right away to see how to locate the part.

Video 9: Locating the Model (Part) through Cognex Pattern Tool

2.7. Cognex In-Sight Part Inspection Tools Part 1

In the last demo we located our tool. We have our "Locate Parts or Pattern" specified for our screwdriver. Now we're going to move on to the inspection part. There are quite a few tools that you can use to locate your part, but I want you to experiment them on your own.

I might make a few more videos in the later section exploring each tool one by one, but essentially, they do pretty much the same thing. You can look at edges, locate edge intersections, blobs, so on and so forth. Let's go to the next demo now to learn the entire process.

Video10: Cognex In-Sight Part Inspection Tools Part 1

2.8. Cognex In-Sight Part Inspection Tools Part 2

In the last demo video, we learned how to create tools which calculate the distance or the length of our tool as well as the width of the handle of a screwdriver. One thing that I did want to point out is that as we added these tools, one thing you may not have noticed is that the time of this application didn't change that much from the first time we ran it.

What really takes the time, which you will see this as we experiment, is finding your model region. You will learn this and much more in this demo:

Video11: Cognex In-Sight Part Inspection Tools Part 2

2.9. External Trigger - Button Setup to Trigger Cognex In-Sight Camera

As you can probably already tell, the setup is quite a bit different than what you've been used to in the last couple of demos. The reason for that is we've reached a critical point out, so I believe it is very important to revisit the hardware once again.

What I mentioned at the beginning of this topic, as well as the hardware one, is that I was only powering the camera and using the bare minimum in order to create a setup and essentially a way to capture an image in order to create the program which we're going to be using.

That being said, what you will not see in a real-world scenario, or on a manufacturing floor, is a trigger which is continuous, which is the setting that we've used a couple of lectures back, and which we've kept throughout the tutorial.

Now we are going to revisit the pinout cable from the camera in order to make the right connections and set ourselves up for success. I'm essentially going to show you the right way to trigger a camera.

One of the proper ways is through the use of push buttons. For example, if you have a simple product that you want to put, inside of the camera's view, or at the beginning of a run, you can use just an ordinary push button with a normally open contact. I'm going to show you how to wire that in.

We're going to test that functionality in insight Explorer, and we're going to see how the camera behaves in that particular scenario. You should treat it as just any other signal, such as a relay, for example, or just an output from a PLC, for example.

But the real way that I've seen it in many plants is what's called a **proximity sensor**, an example of which is shown in Fig. 2.9.1.

Fig. 2.9.1: A proximity sensor

You've probably seen this. It's just a generic sensor. You might have a different use of them in your particular facility, but essentially, it's just a prox which detects a certain product coming through.

It sends back a signal (a 24 volt), either a PNP or NPN signal to the camera. The Cognex 7000 series camera allows you to wire based on both of those two different schemes, and essentially triggers the entire process of capturing the image, processing that data, and then spewing out the result in both scenarios.

Now let's jump right into the demo to have a view of my desk. I'm going to talk about the connections once again, about the pinout cable, and how you need to wire in the hardware.

Video12: External Trigger - Button Setup to Trigger Cognex In-Sight Camera Part 1

2.10. External Trigger - Sensor Setup to Trigger Cognex In-Sight Camera

Welcome to the second demo on triggers. It's going to be a quick one. I just wanted to demonstrate a feature which is the sensor instead of the button. I mentioned that in the previous demo.

The setup is going to be fairly straightforward. We have the same camera and the same setup. As you will see in the demo, the camera's already online and is ready to go. Let's dive right in.

Video13: External Trigger - Sensor Setup to Trigger Cognex In-Sight Camera Part 2

Chapter 3. Cognex Intermediate Topics

In this chapter, we will discuss certain Cognex intermediate topics, such as reading bar codes. But first, if you have been paying attention to the previous demos and for whatever reason you powered down the camera and then powered power it back on, you would have noticed that the camera did not really retain all of the tools that we've created.

3.1. Working & Managing "Job" Files

In this first section, we're going to be addressing the afore-mentioned problem, that is, we will see how to work and manage "job" files. Let's dive right in:

Video 14: Working & Managing Job Files

3.2. Reading 1D Barcodes Job Tutorial Part 1

In this and the next couple of demos we're going to explore a full application, which is reading a barcode of an object. I have a couple of examples, but essentially what I want to test it on is an Allen Bradley IO (CompactLogix input card).

So, we're going to retrieve that barcode and essentially confirm that it's the right barcode through the camera and be able to recognize different barcodes within the same application. In the demo, I'm going to show you step-by-step how to get that done. Let's get started.

Video 15: Reading 1D Barcodes Job Tutorial Part 1

3.3. Reading 1D Barcodes Job Tutorial Part 2

Welcome back to our tutorial on barcode reading. Now we're going to proceed with our application. Since our barcode is almost on the entire screen, we're not necessarily going to click the "Locate Part" button. Let's go to the demo now to see what exactly is to be done.

Video 16: Reading 1D Barcodes Job Tutorial Part 2

3.4. Reading 1D Barcodes Job Tutorial Part 3

We are still on our tutorial on Cognex where we are reading one-dimensional barcodes. In this section I want to explore a different object. It's a barcode on a packaging from a specific brand of chocolate.

We're going to put it under the test and see if our system will not only able to recognize a different barcode, but also be able to tell us that the string is different, and essentially reject this product since it's not the part that we are expecting within our system. Let's go to the demo to begin.

Video 17: Reading 1D Barcodes Job Tutorial Part 3

Chapter 4. Cognex Integration into Studio 5000 Rockwell Environment

This is a very interesting section in which we're going to explore the integration of a vision system on the Cognex In-sight 7200 series into our PLC.

4.1. Initializing Studio 5000 Program & Preliminary Configuration

The camera is going to read a certain barcode and be able to send that string after combining those bytes of information into double integers back to our PLC system.

This is a very important notion in PLC systems nowadays, because at the end of the day, you're not working purely with the inputs and outputs on your PLC, but oftentimes as a controls Engineer/Technician, you have to integrate different systems from different manufacturers into your own system. Now let's go to the demo to begin.

Video 18: Initializing Studio 5000 Program & Preliminary Configuration

4.2. Cognex AOP (Add-On Profile) Installation & Integration

In this section we continue our Cognex integration tutorial. We're going to look at creating a link between the Cognex camera and the CompactLogix processor. This is done over ethernet, that is, both of these devices communicate over Ethernet IP. This is the most common protocol that is used on the plant floor nowadays.

Most of your servo drives have moved away from what's called circus networks down to ethernet IP, or what's called the CIP. Without further delay, let's jump right in to the demo.

Video 19: Cognex AOP (Add-On Profile) Installation & Integration

4.3. Downloading Studio 5000 Program to the PLC & Testing Communication to Module

In our Cognex integration, the next step that is extremely critical is going online with the processor and not only making sure that the program is working okay, but also that we are able to talk with the camera the way it is supposed to. Let's get started.

Video 20: Downloading Studio 5000 Program to the PLC & Testing Communication to Module

4.4. Initializing Cognex Job & Basic Configuration for Barcode 1D Reader

In this section you will learn how to initialize a Cognex job and perform basic configuration. As you will see in the demo, the In-Sight explorer we are using is called EasyBuilder, just to give you a good idea in case you're not familiar with the software.

The version we are using is 5.6.1, build 841. I tell you this is just to make sure that you are comfortable with the version and that you have something similar. Now let's go over to the demo to complete the tutorial.

Video 21: Initializing Cognex Job & Basic Configuration for Barcode 1D Reader

4.5. Cognex Communication to PLC & Setting Up Barcode Data Passing

Now we are reading the barcode of our device, and we are able to see that data on our screen, through the Cognex In-Sights 7200 series. The next step is to establish communications back to our PLC. Let us go to the tutorial now to learn how to do that.

Video 22: Cognex Job Change through EtherNet_IP using a CompactLogix PLC System

4.6. Cognex Job Change through EtherNet/IP using a CompactLogix PLC System

In the last tutorial, we were able to create the string and set up the communications which are going to go back from the Cognex camera into our PLC. It’s a very cool and exciting technology right there. We are going to switch over to the PLC but just before we do that, let's put the camera back online first. Let’s go over to the demo now.

Video 23: Cognex Job Change through EtherNet_IP using a CompactLogix PLC System

Chapter 5. How to Download the Training Videos, PLC Programs (Codes) & Demo Editions of RSLogix 5000 / Studio 5000 Logix Designer

Below are the links to download all the 23 demo videos in this section, PLC programs and the Demo edition of RSLogix 5000 / Studio 5000 Logix Designer:

https://bit.ly/3sxN7jr

When you open this link, you should be able to download:

1. All the 23 training videos/tutorials (in mp4 format)

2. Copyright and License Info (txt file)

3. PLC Program files (codes), titled “My_PLC_Development_Zipped” (Zipped file)

Here’s the link to download any of the **demo versions** of Logix designers and emulators:

Use the following BIT.LY link to download Various Demo Versions of RSLogix 5000 and Emulator Designer or use this short link: http://bit.ly/2lTyw3F.

I use version 20 and version 17 in most of my projects. Version 17 has an expiry time of 90 days. It also has a few limitations, but you will be gaining experience with the real thing. Here is the link to download it:

Use the following BIT.LY link to download RSLogix 5000 Designer V17: http://bit.ly/2lyjosq.

During the installation of Version 17, you will be asked for **serial number**. Just enter a random number such as 123 or 12345. Use the following BIT.LY link to download the full installation guide: https://bit.ly/35VHVg6.

Chapter 6. How to Get Further Help

Dear reader, I hope you find this section and the accompanying videos very helpful in learning how to develop and embed the Cognex In-Sight Machine Vision System in Programmable Logic Controller (PLC) with my training videos. You may find some of the concepts I covered in this book confusing at first. However, with time and a little more effort, you should be able to embed any vision system in a PLC and do other great stuff.

Contact me through my support email below or visit my Author page if you need further help, or if have questions or requests.

Regards,

Charles J. J.

amandamorin3000@gmail.com.

Section 4: How to Integrate & Program Point IO Hardware in RSLogix 5000 with Training Videos

This section is an exhaustive collection of my tutorials and training videos on how to work with Allen Bradley's Point IO 1734 hardware.

First, you will be introduced to this hardware component from Rockwell Automation. Next, you will learn how to connect to it from a network. With my training videos, you will also learn how to wire a SICK sensor into a Point IO input and test the hardware in Studio 5000.

There are 11 chapters in this section which are accompanied with **9 in-depth HD training videos that you can download**. These videos simplify everything you need to understand, and help you **speed up your learning** of Point IO Hardware programming and integration. There is also a link in this book for you to download my PLC programs (codes) for your revision.

I start with an overview of the Point IO modules, and then move on to IP addressing, web server, and power supply overview. Then I proceed to Point IO integration into Studio 5000, including how to create the Point IO definition within the CompactLogix 1769-L24ER-QB1B PLC and validating the communication. Finally, I take a special look at a Point IO real panel example. This is a Point IO panel used for a plant control which I commissioned myself few months ago.

So, after studying this section and the training videos, you should develop a hands-on approach to the programming and integration of Point IO into Studio 5000 Rockwell Environment, and be able to use it in real world industrial applications.

The best way to master PLC programming is to use real world situations. The real-world scenarios and industrial applications developed in this book and its accompanying video trainings will help you learn better and faster many of the functions and features of Studio 5000 platform.

The methods presented in the training videos are those that are usually employed in the real world of industrial automation, and they may be all that you will ever need to learn. The information in this book and the training videos is very valuable, not only to those who are just starting out, but also to any other skillful PLC programmer, no matter their skill level.

Merely having the user manual of a Point IO hardware, or referring to its help contents, is far from enough in learning this critical component of a distributed control system. Therefore, this book and the accompanying demos are extremely useful for learning how to reduce costs of installation and the need of multiple PLCs in many situations.

However, if you are a complete novice to PLCs or PLC programming, here's the link to my book that you will find very helpful to get started: PLC Programming Using RSLogix 500 & Industrial Applications: A Guide for Learning Ladder Logic Concepts Step by Step . Then you can move on to the first section of this book.

You will find the link to download all the 9 training videos and program codes in Chapter 10 of this section. The serial number and title of the video you should watch is mentioned in the appropriate section of each chapter.

For effective learning and to get good practice, I strongly advise you attempt the challenges and projects in this book on your own. Build your own program after you have watched me build mine in each video. Then later you can cross check your programs with mine. Also feel free to use or modify any of the PLC programs you downloaded as you wish.

Finally, if you have questions or need further help, use the support link I provided in Chapter 11. I will get back to you quickly.

Chapter 1. Introduction to Point IO PLC Programming

This chapter is an introduction to the Point IO. A Point IO is a module which helps you to increase the number of inputs and outputs at the machine level. It plays a very critical role within modern and distributed control system environments.

It also helps the end user to add input and output nodes to their system within proximity of field devices. Therefore, it helps to minimize costs of installation and the need of multiple PLCs in many situations.

Other PLCs such as CompactLogix L24ER feature local inputs and outputs. However, the question which arises is what can you do once you start getting more inputs and outputs in the PLC field? The answer is we should instead use remote modules, such as Point IO and Flex IO from Alan Bradley and many others.

Let's go over to the first demo video for your first introduction to the Point IO.

Video 1: Point IO Introduction (2:57 Minutes)

Chapter 2. Hardware Overview of Point IO Module

In this chapter we will look more closely at the Point IO module. In the demo video I have an assembled Point IO module which we're going to take apart. We will go over each and every device, learn how it fits together into the overall infrastructure, and also know what kind of decisions you may want to make in the system.

So, let's go to the demo to look at the way these modules communicate:

Video 2: Point IO Hardware Overview (9:24 Minutes)

Chapter 3. How to Power on the Point IO Module

In this demo on Point IO, we're going to power on our module. We will then go on to discuss the different cards, my experience with them and what I would recommend depending on the choice of inputs and outputs for your specific system.

So, without any further delay, let's take a look at the power on.

Video 3: Point IO Modules (12:33 Minutes)

Chapter 4. EtherNet IP Address Configuration BOOTP & Thumbwheel

Now that we are familiar with the different Point IO building blocks, as well as the different input and output cards, it is time for us to connect to this module from the network as well as from the PLC.

Video 4: Point IO IP Addressing (08:53 Minutes)

Chapter 5. Exploring the Web Connectivity of the Point IO Rack

In this demo on Point IO, we're going to explore the web connectivity, which the Point IO rack has to our computer. This is extremely important for setup purposes, troubleshooting purposes, or for just identifying the different modules in the field.

So, let's go back to the IP address to check this out.

Video 5: Point IO Web Server and Power Supply Overview (08:40 Minutes)

Chapter 6. Point IO Integration: How to Create Point IO Definition Within CompactLogix PLC & Validate the Communication

Welcome to this demo on Point IO. As you will see in the video, we have the Point IO rack ready, powered on, and good to go. As far as the cards are concerned, there are a couple of errors that you will notice.

There's a blinking red light. Also, some network status lights are not lit up. For example, there are some network status lights on two cards on the far right that do not light up. So, we're going to be solving these problems in the demo video.

Video 6: Point IO Integration into Studio 5000 - Module and Tag Overview (13:38 Minutes)

Chapter 7. Point IO Digital Input Wiring & Testing: How to Wire a Digital SICK Sensor into a Point IO Input & Testing the Hardware in Studio 5000

In this demo, we continue our discussion on Point IO. We will use the same rack that we configured in the previous demo., you will notice, however, a couple of differences since the beginning of the previous demo.

Now, we're going to wire in a simple sensor, just for verifying that everything on the Point IO is working. Without any further delay let's get started.

Video 7: Point IO Sensor Wiring & Testing Logic - Module and Tag Overview (10:29 Minutes)

Chapter 8. Point IO Programming: How to Create PLC Based Tags for Point IO Based Inputs & Outputs

Welcome to another tutorial on point IO in which we're going to be focusing on programming the sensor which we've plugged in the previous demo. Now we're going to program the inputs and outputs on a rack correctly so that they work in our PLC so that we have access to the proper tags, as well as the proper structure. Without any further delay let's get started.

Video 8: Point IO Input & Output Programming Routines - Module & Tag Overview (12:12 Minutes)

Chapter 9. Point IO Practical Example: A Real Panel Used for Plant Control

This chapter is a practical demo on Point IO in which we're going to discuss a field-installed Point IO rack. This Point IO is actually inside of a panel which I commissioned about six months ago in a plant that needed remote access points.

There's going to be a main control system, which hosts a ControlLogix PLC that is able to communicate with this panel and get the inputs and outputs as they were needed within the remote area.

Let's look at it now:

Video 9: Point IO Practical Example - Module & Tag Overview (07:49 Minutes)

Chapter 10. How to Download the Training Videos, PLC Programs (Codes) & Demo Editions of RSLogix 5000 / Studio 5000 Logix Designer

Below are the links to download all the 9 training videos, PLC programs and the Demo edition of RSLogix 5000 / Studio 5000 Logix Designer:

https://bit.ly/3y1eL9Z – When you open this BIT.LY link, you should be able to download:

1. All the 9 training videos/tutorials (in mp4 format)

2. Copyright and License Info (txt file)

3. PLC Program files (codes), titled “My_PLC_Development_Zipped” (Zipped file)

Here’s the link to download any of the **demo versions** of Logix designers and emulators:

Use the following BIT.LY link to download Various Demo Versions of RSLogix 5000 and Emulator Designer: http://bit.ly/2lTyw3F.

I use version 20 and version 17 in most of my projects. Version 17 has an expiry time of 90 days. It also has a few limitations, but you will be gaining experience with the real thing. Here is the link to download it:

Use the following BIT.LY link to download RSLogix 5000 Designer V17: http://bit.ly/2lyjosq.

During the installation of Version 17, you will be asked for **serial number**. Just enter a random number such as 123 or 12345. Use the following BIT.LY link to download the full installation guide: https://bit.ly/35VHVg6.

Chapter 11. How to Get Further Help

Dear reader, I hope you find this section of the book and the accompanying videos very helpful in learning how to integrate and program Point IO Hardware in RSLogix 5000 with my training videos.

You may find some of the concepts I covered in this section and the rest of the book confusing at first. However, with time and a little more effort, you should be able to integrate and program Point IO Hardware in RSLogix 5000 and do other great stuff discussed in the rest of the book.

Contact me through my support email below or visit my Author page if you need further help, or if have questions or requests.

Cheers,

Charles J. J.

Amandamorin3000@gmail.com

More Helpful Resources

You may find the following books helpful. They are published by Ojula Technology Innovations, and are available at all major online book retailers:

1. **Cybersecurity Essentials**: The Beginner's Guide.

2. **Cloud Computing Fundamentals**: Learn the Latest Cloud Technology and Architecture with Real-World Examples and Applications.

3. **AWS Certified Cloud Practitioner Exam Success Guides, Volumes 1 & 2**: Prepare for CLF-C01 Exam with quizzes, assessment tests, cheat sheets, hands-on lessons and practice exams to build real-life experiences.

4. **Web Development from Beginner to Paid Professional, Volumes 1 & 2**: Build your portfolio as you learn Html5, CSS and Javascript step by step with support.

5. **Google Cloud Certified Associate Cloud Engineer Certification Guide**, Volume 1: Learn with challenge labs, assessment tests and practice exams to build innovations and real-life experiences.

6. **Practical Electronic Circuits:** A Strong Foundation for Creating Electronic Projects.

Made in United States
Orlando, FL
04 November 2026

90206127R00069

Made in United States
Orlando, FL
04 November 2026

90206127R00069